Single-Event Effects, from Space to Accelerator Environments

Ygor Quadros de Aguiar • Frédéric Wrobel •
Jean-Luc Autran • Rubén García Alía

Single-Event Effects, from Space to Accelerator Environments

Analysis, Prediction and Hardening by Design

 Springer

Ygor Quadros de Aguiar
European Organization for Nuclear
Research (CERN)
Geneva, Switzerland

Jean-Luc Autran
University of Rennes
Rennes, France

Frédéric Wrobel
University of Montpellier
Montpellier, France

Rubén García Alía
CERN–European Organization for Nuclear
Research (CERN)
Geneva, Switzerland

ISBN 978-3-031-71722-2   ISBN 978-3-031-71723-9   (eBook)
https://doi.org/10.1007/978-3-031-71723-9

This work was supported by CERN

This Springer imprint is published by the registered company Springer Nature Switzerland AG
The registered company address is: Gewerbestrasse 11, 6330 Cham, Switzerland

If disposing of this product, please recycle the paper.

Preface

As humanity pushes the boundaries of exploration and technology, the need to understand and mitigate the effects of radiation on electronic components in harsh environments, such as space missions and particle accelerators, has never been more critical. This book addresses that need by providing a thorough examination of radiation-induced failure mechanisms, with a particular focus on Single-Event Effects (SEEs) that pose significant challenges to the reliability of modern electronics, and how to mitigate them at the design level.

For newcomers to this field, this book offers a solid foundation in the fundamental concepts of particle interaction physics and electronics hardening design. We begin with the basics-the composition and dynamics of radiation environments and their impact on electronic components. From there, we explore the principles of component qualification and the design techniques used to harden electronics against radiation-induced failures. These sections are designed to equip new researchers and engineers with the knowledge they need to navigate this complex field.

For more experienced readers, this book provides a comprehensive discussion of the state-of-the-art methodologies in modeling and advanced Radiation-Hardening-by-Design (RHBD) techniques that can be employed at both the physical layout and circuit levels. The layout design of integrated circuits plays a significant role in influencing SEE generation, and thus, layout-level hardening techniques are critical for minimizing radiation-induced charge collection. At the circuit level, this book explores the implications of logic synthesis in cell-based designs, proposing signal probability as a novel, application-specific approach to hardening. This method promises to enhance the efficiency of RHBD techniques while minimizing design drawbacks and avoiding misleading qualifications.

One of the key insights presented in this book is the potential for optimization in RHBD techniques. For instance, optimizing pin assignment to target transient effects can reduce the overall failure rate without increasing the design area. Similarly, selective Triple-Modular Redundancy (TMR) block insertion methodologies can be refined based on the signal probability of critical nodes and the robustness of majority voter architectures, leading to more efficient and reliable designs.

This book is the culmination of the extensive research carried out during my PhD thesis, developed under the RADSAGA Marie Curie fellowship. It has been a privilege to contribute to this rapidly evolving field alongside esteemed colleagues and mentors, whose expertise and guidance have been invaluable. I hope that this work will serve as a useful resource for those seeking to understand and mitigate the challenges posed by radiation-induced failures in electronic components.

Finally, I would like to express my deepest gratitude to my co-authors-Professors Frédéric Wrobel, Jean-Luc Autran, and Dr. Rubén García Alía-whose contributions and insights have significantly enriched this book. Their dedication to advancing our understanding of radiation effects in electronics is truly inspiring, and I am honored to have collaborated with such distinguished experts in the field.

Geneva, Switzerland   Ygor Quadros de Aguiar
08/2024

Acknowledgments

I am deeply grateful to my parents for their support throughout my journey from college to this moment. Without their love, encouragement, and sacrifices, I wouldn't be where I am today. I also extend my heartfelt gratitude to my twin brother and little sister for their unwavering support and encouragement.

Additionally, I wish to extend my appreciation to my colleagues from the RADIAC team at the Université de Montpellier for their friendship and support during my PhD journey. Special appreciation goes to my esteemed colleagues from the Radiation to Electronics (R2E) project at European Organization for Nuclear Research (CERN). Working alongside them has been a privilege, and I am grateful for the opportunities to collaborate and learn from each of them. In particular, I would like to express my sincere thanks to Matteo Ferrari, Samuel Niang, and Kacper Bilko for their friendship, numerous hours of conversations, and the several coffee breaks we shared together.

I owe a debt of gratitude to my supervisors and co-authors, Frédéric Wrobel, Jean-Luc Autran, and Rubén García Alía, for their unwavering support, belief in my potential, and invaluable mentorship over the past few years. Last but not least, I am thankful to God for granting me the strength and courage to pursue my dreams.

Contents

1 Radiation Environment and Their Effects on Electronics 1
  1.1   Introduction 1
  1.2   Radiation Environments 1
       1.2.1   Space Environment 1
       1.2.2   Atmospheric Environment 6
       1.2.3   Accelerator Environment 8
  1.3   Particle Interaction Mechanisms 12
  1.4   Energy Deposition 15
  1.5   Charge Collection 21
       1.5.1   Charge Sharing and Pulse Quenching Effect 23
  1.6   Summary 24
  References 26

2 Introduction to Single-Event Effects 29
  2.1   Context and Overview 29
  2.2   Single-Event Upset (SEU) 31
  2.3   Single-Event Functional Interruption (SEFI) 35
  2.4   Single-Event Transient (SET) 36
       2.4.1   Logical Masking Effect 38
       2.4.2   Electrical Masking Effect 38
       2.4.3   Latching-Window Masking Effect 39
  2.5   Single-Event Latchup (SEL) 40
  2.6   Summary 43
  References 44

3 Single-Event Effect Prediction Methodologies 49
  3.1   Modeling and Prediction Tools 49
  3.2   SEE Triggering Criterion 50
       3.2.1   Rectangular Parallelepiped (RPP) Criterion 50
       3.2.2   Drift-Diffusion-Collection Criterion 52

3.3   Modeling Radiation-Induced Currents in Circuit Simulations   53
3.4   Proposed Prediction Methodology Based on the Diffusion Model ...   55
3.5   Summary   58
References   59

4 Radiation Hardening   63
4.1   Introduction   63
4.2   Radiation Hardening by Process (RHBP)..........................   65
4.3   Radiation Hardening by Design (RHBD)...........................   68
    4.3.1   Layout-Based Techniques   68
    4.3.2   Circuit-Based Techniques   72
4.4   Summary   76
References   77

5 Analysis of Layout-Based RHBD Techniques   81
5.1   Introduction   81
5.2   Part I—Gate Sizing and Transistor Stacking   82
    5.2.1   Gate Sizing (GS)...   82
    5.2.2   Transistor Stacking (TS)..   86
    5.2.3   Comparison of Power and Area Overhead   88
    5.2.4   Impact on the SET Cross Section   89
5.3   Part II—Transistor Folding (TF) and Diffusion Splitting (DS)   94
    5.3.1   Impact on the SET Cross Section   97
    5.3.2   Asymmetric Designs.. 101
    5.3.3   Voltage Variability 104
    5.3.4   Impact on the In-Orbit SET Rate: LEO and ISS Orbits....... 106
    5.3.5   Transistor Scaling and Angular Dependence 107
5.4   Summary 108
References 110

6 Analysis of Circuit-Based RHBD Techniques 115
6.1   Reliability-Driven Synthesis.. 115
    6.1.1   Multiple V_{th} Cells and Voltage Scaling...................... 118
    6.1.2   Technology Mapping 120
6.2   Pin Assignment 125
    6.2.1   Optimization of Pin Assignment for Single-Event
            Transients 125
    6.2.2   Impact on the SET Cross Section of Standard Cells 129
6.3   Summary 133
References 134

Index 139

Chapter 1
Radiation Environment and Their Effects on Electronics

1.1 Introduction

From the terrestrial landscapes of Earth to the celestial expanses of our solar system, ionizing radiation is omnipresent. It shapes the world around us, playing an indispensable role in sustaining life on Earth while also posing potential threats. Radiation itself is a form of energy emitted by atoms, taking the form of either electromagnetic waves or energetic particles. Different types of radiation interact with matter in unique ways, and they can be generated naturally, as in the case of galactic cosmic rays, or artificially, as in particle accelerators. Furthermore, the intensity and their energy vary significantly across environments. Therefore, understanding the sources and characteristics of radiation within a specific environment is crucial for ensuring safety and the effective operation of electronic systems in any given environment.

1.2 Radiation Environments

1.2.1 Space Environment

Mission success in the near-Earth space environment hinges on a thorough understanding of the prevalent radiation sources (as illustrated in Fig. 1.1). These sources can be broadly categorized into three primary types:

- **Solar Energetic Particles (SEPs):** continuously emitted by the Sun during periods of enhanced activity, SEPs are energetic protons and heavier ions accelerated by solar flares and coronal mass ejections (CMEs). The intensity and energy spectrum of SEPs vary significantly depending on the specific solar event.

© The Author(s) 2025
Y. Quadros de Aguiar et al., *Single-Event Effects, from Space to Accelerator Environments*, https://doi.org/10.1007/978-3-031-71723-9_1

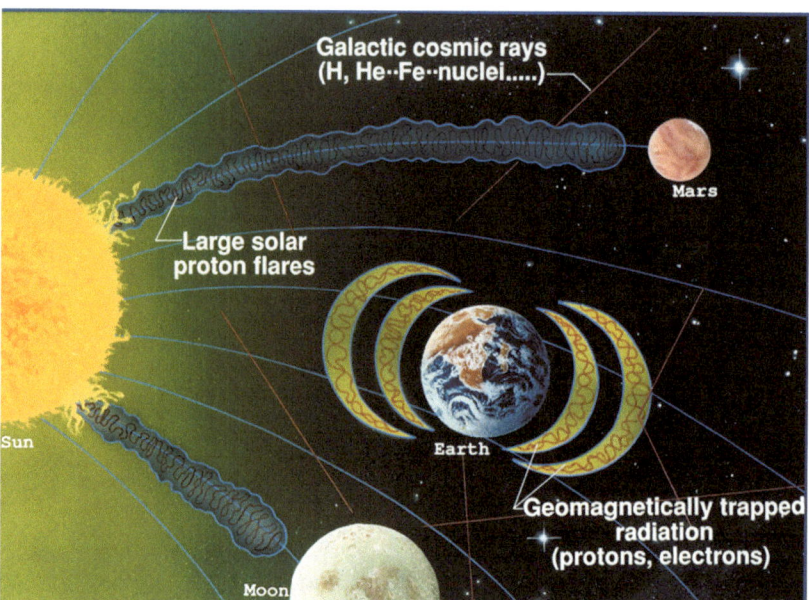

Fig. 1.1 A schematic diagram illustrating the three primary sources of radiation in near-Earth space (Adapted from [1])

- **Galactic Cosmic Rays (GCRs):** originating from outside our solar system, GCRs are high-energy particles, primarily consisting of protons and heavier nuclei, believed to be produced by violent events such as supernovae explosions. Due to their high energy, GCRs can penetrate deep into spacecraft and pose significant health risks to astronauts on long-duration missions. GCRs are a constant source of radiation in near-Earth space, but their flux is modulated by the Sun's heliosphere, a vast bubble of charged particles extending outward from the Sun.
- **Geomagnetically Trapped Particle Radiation (Van Allen Belts):** this region consists of a complex mixture of energetic particles, primarily protons and electrons, trapped by the Earth's magnetosphere. The magnetosphere is a vast, dynamic region shaped by the Earth's magnetic field, which acts as a shield against incoming charged particles from the Sun and beyond. These trapped particles form two distinct toroidal regions known as the Van Allen radiation belts. The inner belt primarily consists of high-energy protons, while the outer belt contains a broader mix of protons and electrons. The specific location and intensity of trapped particles within these belts vary depending on space weather conditions.

Solar activity plays a crucial role in shaping the near-Earth space radiation environment. Solar flares and CMEs can significantly enhance the flux of SEPs, potentially creating intense radiation bursts that can overwhelm spacecraft shielding. Furthermore, solar activity can also indirectly influence the intensity of GCRs and trapped radiation. During periods of high solar activity, the Earth's magneto-sphere becomes more compressed, allowing a higher flux of GCRs to penetrate into the inner regions near-Earth space. Additionally, energetic solar wind particles can interact with the magnetosphere, leading to enhanced particle populations within the Van Allen belts.

Therefore, understanding and monitoring solar activity through observations of sunspots and other solar phenomena is critical for predicting the radiation environment that spacecraft and sensitive electronics may encounter in near-Earth space.

Particle flux is the rate of particle that passes through a unit area per unit time, expressed in $cm^{-2} \cdot s^{-1}$, also known as *flux density*. Another widely used quantity is the **particle fluence**, which is the particle flux integrated over a period of time and expressed in cm^{-2}.

Numerous space missions have been specifically designed to measure and comprehend solar activity, recognizing its significance for life on Earth and its influence on mission planning and design. A notable example is the Solar and Heliospheric Observatory (SOHO), a joint mission between the European Space Agency (ESA) and NASA, which was launched in 1995 and continues to operate successfully.

The Sun's activity follows an approximate 11-year cycle, characterized by periods of high (solar maximum, about 7 years) and low (solar minimum, about 4 years) activity. Figure 1.2 presents solar activity measurements and predictions for the upcoming cycles based on the sunspot number, also known as the *Wolf number*.

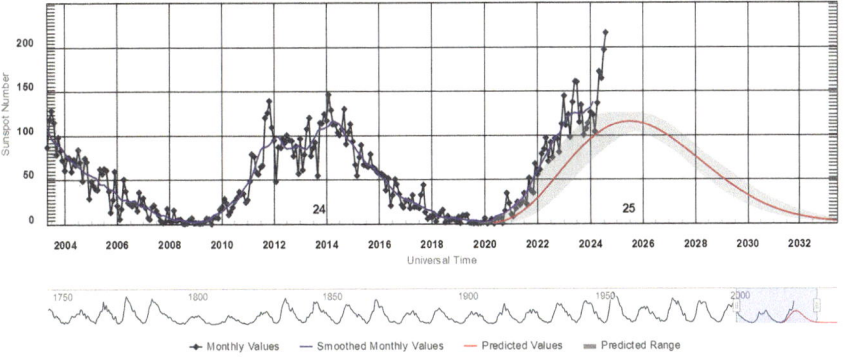

Fig. 1.2 Solar Cycle Sunspot Number Progression. The peak of cycle 25 is expected in 2025 [4]

Sunspots are cooler, darker regions on the Sun's surface associated with intense magnetic activity.

While the Sun continuously emits a low-energy stream of particles known as the solar wind, primarily consisting of protons and electrons with some heavier ions, significant radiation events are associated with solar flares and coronal mass ejections (CMEs). These infrequent but powerful events release a substantial number of highly energetic particles into space. Solar wind particles typically have energies ranging from kiloelectronvolts (keV) to gigaelectronvolts (GeV) and can reach speeds up to 80% of the speed of light.

Highly energetic particles originating from intense solar activities are particularly concerning due to their ability to rapidly reach Earth's atmosphere within hours or days. These energetic particles pose a dual threat: endangering astronauts and disrupting electronic systems in space [2]. The intensity of the solar wind and the frequency of solar particle events are directly influenced by solar activity, specifically the number of sunspots. During the solar maximum, particularly the declining phase, CMEs and solar flares occur more frequently [3].

Interestingly, solar activity has an inverse correlation with the flux of galactic cosmic rays (GCRs). GCRs are high-energy particles originating from outside the solar system and have the lowest particle flux among the radiation sources in near-Earth space. However, due to their high energy, they are highly penetrating and pose a significant threat to space-borne electronics and human health. Shielding technology offers limited protection against GCR radiation exposure.

During periods of high solar activity, the Sun's magnetic field strengthens, acting as a more effective shield against incoming particles from beyond the solar system. This leads to a decrease in the flux of GCRs reaching Earth's vicinity [5–7]. Conversely, during solar minimum periods, the weakened solar magnetic field allows a higher GCR flux to penetrate into near-Earth space. Understanding and monitoring solar activity and its associated radiation environment is crucial for several reasons:

- **Space Mission Planning:** knowledge of the radiation environment allows engineers to design spacecraft with appropriate shielding strategies and operational procedures to ensure the safety and functionality of spacecraft systems.
- **Astronaut Safety:** astronauts on long-duration missions are exposed to elevated levels of radiation, posing potential health risks. Understanding the radiation environment helps to develop strategies for mitigating these risks.
- **Space Weather Forecasting:** missions like SOHO provide valuable data that contributes to improved space weather forecasting and radiation risk assessment.

Solar events, such as solar storms and CMEs, have been documented for centuries. The *Carrington event* of 1859 serves as a powerful historical example. This massive solar storm caused significant disruptions to early telegraph systems and produced auroras visible across the globe [8, 9]. It was one of the first instances where the consequences of a solar event affecting Earth were recognized. One event in this level could be very catastrophic nowadays as modern technology is completely dependent on electricity. A more recent example occurred in January

2022, where a significant solar storm originating from a CME caused unexpected impacts on orbiting satellites [10, 11]. The solar energy from the CME increased atmospheric drag, pulling several satellites out of their intended orbits. This incident, unfortunately resulting in the loss of multiple satellites, underscores the importance of considering solar activity's potential effects when designing space missions.

The third radiation component in the near-Earth environment is the trapped particle radiation in the Van Allen radiation belts, illustrated in Fig. 1.3. As can be seen in Fig. 1.3, the outer belt is wider than the inner belt, and it is also the most unstable due to the weaker influence of the Earth's magnetosphere. Similar to the GCRs, the trapped radiation suffers influence from the solar activity, and it is modulated such that the higher the activity, the higher is the electron flux and the lower the proton flux and vice versa [12]. Van Allen radiation belts have been always a concern for space mission designs due to their impact on electronic reliability. Protons are able to induce a variety of effects ranging from parametric degradation due to absorbed dose to even singular effects.

These radiation effects are further detailed in the next section. The analysis of the proton flux in the inner belt is highly important for any space mission targeting low-Earth orbit (LEO) as, for example, several Earth observation instruments and the International Space Station (ISS) shown in Fig. 1.3. Overall, a comprehensive understanding of the dynamics and composition of the trapped radiation belts plays a crucial role in developing effective measures to safeguard satellites against the adverse impacts of space weather.

Another crucial consideration in the design of spaceborne components is the anomaly within the Earth's magnetosphere, known as the *South Atlantic Anomaly*

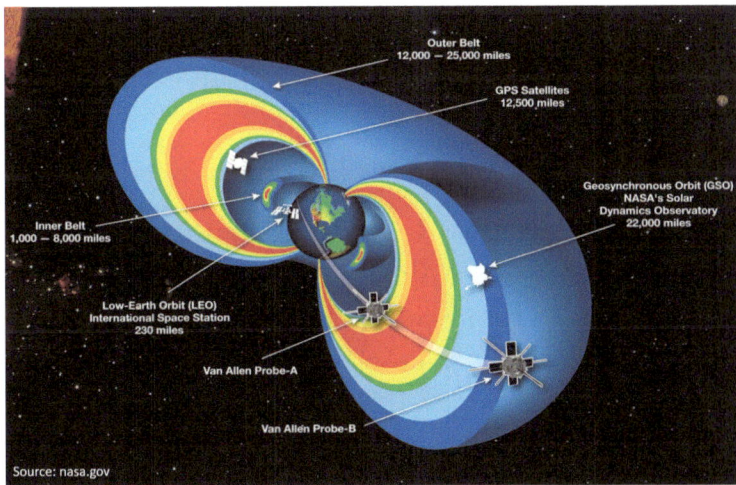

Fig. 1.3 Van Allen Radiation Belts [1]. The outer belt is predominantly composed of electrons, while the inner belt is predominantly composed of protons

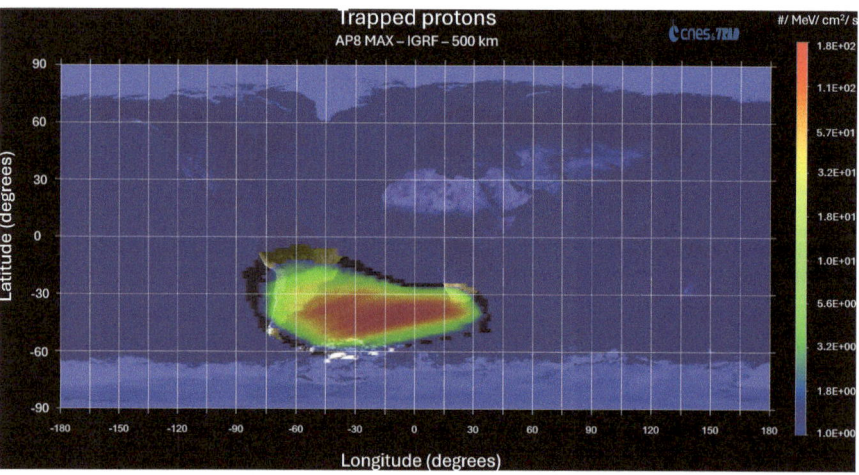

Fig. 1.4 Flux intensity map for the >10 MeV channel at 500 km altitude. Higher proton flux in the South Atlantic Anomaly (SAA) region. The graph was obtained using the AP8 model available in the OMERE software tool [13]

(SAA). This anomaly arises from a slight tilt and offset between the Earth's geomagnetic and rotational axes, causing a significant weakening of the Earth's magnetic field over the South Atlantic region. This weakened field allows energetic particles from the Van Allen belts and galactic cosmic rays (GCRs) to penetrate lower altitudes than they typically would elsewhere, resulting in a zone of increased radiation exposure. Figure 1.4 illustrates this phenomenon, showing the AP8 MIN model's depiction of high-flux protons reaching altitudes as low as 500 km and below in the SAA region.

The presence of the SAA poses a significant challenge for spacecraft operating in LEO. The intense radiation environment within the SAA can potentially damage sensitive electronics onboard satellites. As a result, many satellites employ a precautionary measure of switching off sensitive electronic components when traversing the SAA to avoid potential damage, particularly from stochastic events (discussed in the next chapter).

1.2.2 Atmospheric Environment

When cosmic rays enter the Earth's atmosphere, they interact with atmospheric atoms, primarily nitrogen and oxygen molecules, leading to the production of secondary radiation. Similarly, these secondary particles can also interact with atmospheric atoms, creating a cascade effect and generating additional secondary particles. This phenomenon is known as a *cosmic ray air shower*, and it is illustrated in Fig. 1.5. In this example, the primary cosmic particle is a highly energetic proton

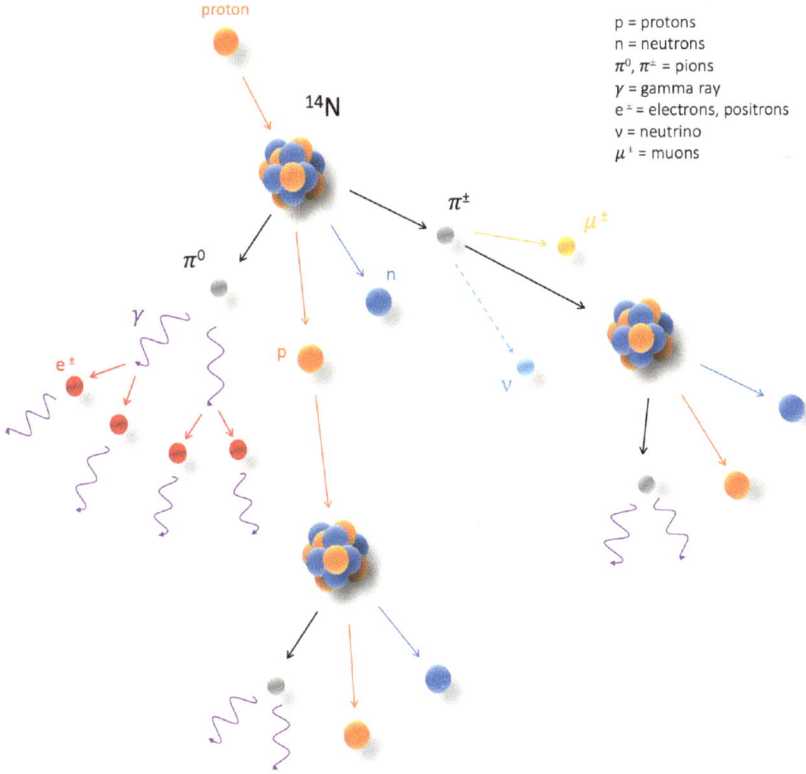

Fig. 1.5 Secondary radiation induced by a high-energetic proton interacting with an atmospheric atom, nitrogen (^{14}N), and leading to a cascade of secondary interactions. Note that this representation provides a simplified illustration of some possible reactions rather than an exhaustive account of all interactions

that collides with the nuclei of a nitrogen atom in the air. These collisions lead to the production of secondary particles, including protons, neutrons, pions π, kaons K, and muons μ. This process continues until the secondary particles have insufficient energy to sustain further interactions. Pions and kaons are unstable particles with short lifespans. They rapidly decay into various particles, including muons, electrons e^-, positron e^+ (antiparticle of electrons), neutrinos ν, and photons γ.

Cosmic ray air showers are comprised of various secondary particles, each with unique properties. Muons are particularly noteworthy due to their relatively long lifespan compared to pions and kaons, despite still being short-lived compared to stable particles like protons and electrons. Although their average lifetime is just 2.2 microseconds; this allows them to travel several kilometers through the atmosphere before decaying. Similar to electrons, muons are fundamental particles. However, their much greater mass grants them significantly higher penetration power. Muons can traverse several centimeters of materials like silicon, potentially reaching sensitive electronic components within spacecraft or even electronics on the ground. Historically, secondary neutrons were considered the primary contributors of radiation damage in electronic devices exposed to the atmospheric environment. However, with advancements in miniaturization and increased device sensitivity, muons pose a growing threat to system reliability by inducing radiation damage in microelectronics [14–17].

It is important to note that the cascade of secondary interactions triggered by a high-energy cosmic ray is a complex process with numerous possible reaction pathways. These pathways depend heavily on the specific energy, angle of incidence, and characteristics of the primary particle, along with the properties of the target atoms it interacts with. While Fig. 1.5 provides a simplified illustration, the true nature of these interactions involves a vast array of potential reactions. The flux and composition of atmospheric radiation are subject to fluctuation, influenced by various factors. Solar activity notably impacts the influx of particles bombarding the upper atmosphere, while the Earth's nonuniform magnetic field results in spatial variations in radiation intensity and composition. Regions with weaker magnetic fields, such as the South Atlantic Anomaly discussed earlier, experience heightened fluxes of energetic particles.

Altitude also plays a pivotal role. As we descend through the atmosphere, particle flux initially rises due to collisions with atmospheric nuclei, generating secondary particles. This increase peaks at the *Pfotzer maximum*, after which flux diminishes due to energy loss through interactions, absorption by atmospheric molecules, and decay of unstable particles [18]. For instance, neutron flux increases with decreasing altitude, reaching its peak density of approximately 10^4 neutrons per square centimeter per hour ($n/cm^2/h$) at around 20 km [19, 20]. However, this density drastically declines to about 20 $n/cm^2/h$ at ground level, a mere fraction of its peak value, approximately 0.2%. Apart from neutrons and muons, protons are also a concern for ground-level atmospheric applications, although typically exhibiting lower fluxes when compared to energies up to 400 MeV [21].

While the probability of failure induced by secondary cosmic ray particles is relatively low at ground level, it can still be a concern for safety-critical systems that rely heavily on electronics. These systems require extremely high reliability, and even a small chance of failure can have significant consequences. Therefore, understanding the particle flux and composition at different altitudes and locations is crucial for assessing the radiation threat and designing reliable systems for atmospheric applications.

1.2.3   Accelerator Environment

Another hostile environment with a very particular radiation field is encountered in particle accelerators used in high-energy physics research laboratories. A very good example is the world's largest collider, the large hadron collider (LHC) in the European Organization for Nuclear Research (CERN) in which, in its current

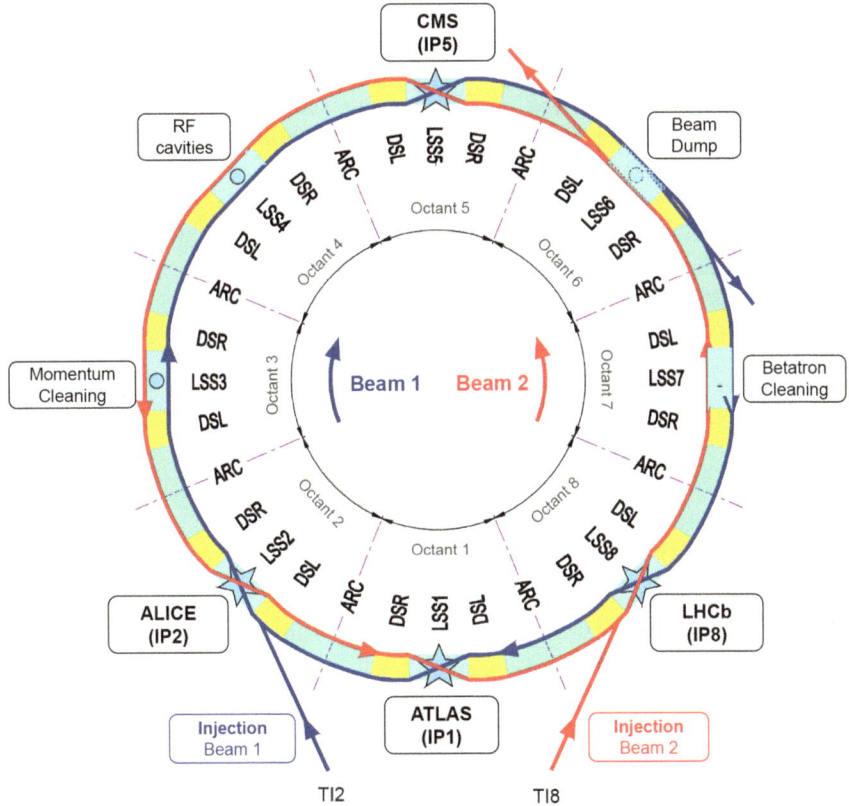

Fig. 1.6 Schematic of the LHC layout division into octants where the main experiments are hosted in IP1 (ATLAS), IP2 (ALICE), IP5 (CMS) and IP8 (LHCb) (Adapted from [22])

design, protons can be accelerated up to 7 TeV. The LHC is a circular collider as shown in the layout in Fig. 1.6 where the eight octants, the interaction points (IP), and the direction of the beams are illustrated. There are four main interaction points where the physics experiments are hosted: interaction point 1 (IP1) hosts the *ATLAS experiment*, the general-purpose LHC experiment where the world's largest particle detector is installed; interaction point 2 (IP2) hosts the *ALICE experiment*, a large ion collider detector dedicated to heavy-ion physics; interaction point 5 (IP5) hosts the compact muon solenoid *(CMS) experiment*, which is another general-purpose detector similar to ATLAS with a broad physics program and a different magnet system; and, lastly, the interaction point 8 (IP8) hosts the *LHCb experiment*, in which, different from the previous experiments, the large hadron collider beauty is a collection of subdetectors used to detect the forward particles from the collisions and to investigate the beauty quark (b quark).

The performance of particle colliders such as the LHC is mostly associated with its physics production, i.e., the number of particle collisions and the production of new physics. Generally, it is measured by the energy that the injected particles can be accelerated to and by the *integrated luminosity* (number of observed inelastic collisions) measured by their detectors. During the LHC physics run 2 (2015–2018), an average integrated luminosity of 50 fb^{-1} (femtobarns) per year was achieved. This high luminosity allowed for a significant number of collision events to be recorded and analyzed. However, with the high luminosity (HL) LHC upgrade project, the nominal annual integrated luminosity is expected to reach over 250 fb^{-1}. The increase in integrated luminosity and the corresponding increase in the number of collisions also lead to higher radiation levels in different areas of the accelerator. These elevated radiation levels can pose operational challenges, such as reduced beam availability caused by beam dumps resulting from magnet quenching (loss of superconductivity) or system failures due to radiation effects on electronics.

> **Luminosity** is a measurement of how many collisions a particle collider can produce. It is usually expressed as an integral over the physics run duration and the unit is the *inverse femtobarn* (fb^{-1}) corresponding to roughly 10^{14} particle interactions at TeV energies [23].

The LHC radiation environment consists of a mixed field with a variety of particles and energies which vary along the different locations in the accelerator complex. It depends not only on the different beam loss mechanisms at play but also on the accelerator settings such as the beam intensity and collimator settings [23, 24]. In this context, there are three main sources of radiation generally related to beam losses:

1. **Beam Interaction with Residual Gas:** the particles are accelerated in a vacuum pipe to prevent their interaction with air molecules and therefore the beam intensity loss. However, despite the very high vacuum levels, there is still some remaining (residual) gas in the vacuum chamber. When the protons interact with these residual molecules, a radiation cascade of secondary particles similar to the one observed in the atmospheric environment is also observed in the accelerator tunnels. This beam loss mechanism is the main contributor to the radiation level in the arc sections in the LHC ring, and it is expected to scale linearly with the *integrated intensity* (number of injected particles) and the residual gas density in the vacuum chamber [25].

2. **Beam Interaction with Machine Elements:** despite the strong magnetic field applied to control and guide the bunch of particles within the beam pipes, some particles, known as *halo particles*, can deviate from the desired trajectory and interact with different beam line elements leading to their activation and increased radiation levels. For instance, to prevent the damage of sensitive components and possible magnet quenches and to provide a protection against uncontrolled beam losses, collimators are used to clean up the beam by absorbing the halo particles. However, the collimator locations are hotspots for radiation levels as a large number of secondary particles is produced from these halo collisions.

3. **Debris from Collisions in the physics experiments:** another common source of radiation is the collision debris from the proton-proton inelastic interactions in the four main LHC experiments shown in Fig. 1.6. For every single proton-proton collision, an average multiplicity of 120 secondary particles are generated and intercepted by the detectors placed in the experiments [23]. However, some of these secondaries are able to reach out of the detector zone and collide with the beam elements in the surrounding area, increasing the radiation levels. Different from the beam losses due to the beam-gas interactions, the radiation levels induced by collision debris are luminosity driven and therefore scale with the integrated luminosity of each experiment [25].

The monitoring of radiation levels in particle accelerators serves multiple crucial purposes. It not only safeguards the integrity and performance of the accelerator itself but also ensures the safety of personnel working in the underground beam lines. While delving into the intricacies of radiation protection measures falls beyond the scope of this book, it is essential to acknowledge their significance in such radioactive environments. Additionally, radiation hardness assurance (RHA) methodologies play a pivotal role in qualifying components and systems to withstand the harsh radiation environment, mitigating the risk of system failures induced by high radiation levels [23, 26–28]. When examining the impact of radiation on electronics, particular attention is given to high-energy hadrons (HEH) and neutrons, as they dominate the particle spectrum within the LHC environment. Figure 1.7 presents a representative lethargy spectrum encountered at the LHC, considering the relevant hadrons such as protons, charged pions, and kaons, as well as neutrons. This

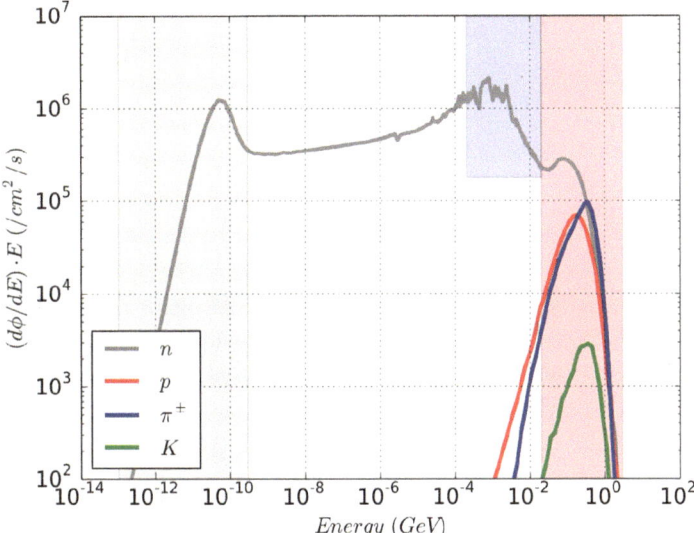

Fig. 1.7 FLUKA-simulated lethargy spectra for an LHC-like test location at the CHARM test facility at CERN. The different shaded regions approximately represent the thermal neutron (gray), intermediate neutron (blue), and HEH fluxes (red) [32]

Table 1.1 Approximate
high-energy hadron (HEH)
annual fluxes from space to
accelerator radiation
environments

Spectrum	Φ HEH (/cm^2/yr)
Ground level	$\sim 2 \cdot 10^5$
Avionics	$\sim 2 \cdot 10^7$
ISS orbit	$\sim 7 \cdot 10^8$
LEO orbit (800 km)	$\sim 3 \cdot 10^9$
LHC ring + transfer lines	$\sim 10^6 - 10^{12}$
Super proton synchrotron (SPS)	$\sim 10^6 - 10^{12}$
Proton synchrotron (PS)	$\sim 10^{10} - 10^{11}$
PS booster	$\sim 10^9 - 10^{11}$

spectrum was simulated using FLUKA, a widely validated Monte Carlo transport code extensively utilized in accelerator research and beyond [29–31]. As depicted in Fig. 1.7, three distinct energy ranges stand out, with varying abundances and effects on electronic reliability. High-energy hadrons exhibit a prominent flux peak in the several hundred MeV range, while neutrons span a range from thermal energies up to 20 MeV [32]. These energy ranges are of significant relevance when assessing the potential impact of radiation on electronic systems. By comprehending the characteristics and intensities of these particles, strategies can be devised to harden electronics against radiation-induced challenges and ensure their reliable operation in the demanding environment of particle accelerators [27].

Hadrons are subatomic particles composed of two or more quarks held together by the strong nuclear force, such as protons, neutrons, pions, and kaons. The term high-energy hadrons (HEH) is widely used in the accelerator environment to design the hadrons with energies in the MeV, which are quite relevant in regard to inducing radiation events on electronic components.

In Table 1.1, the approximate HEH annual fluxes are presented for the different radiation environments, from space, i.e., the International Space Station (ISS) and low-Earth orbits, to ground and accelerator environments. For instance, the HEH fluxes at CERN's accelerator complex, from the injector chain to the LHC ring, can be comparable to the fluxes observed in commercial flights (around 10^7/cm^2/yr) and in space ($10^8 - 10^9$/cm^2/yr); however, depending on the location in machine, it can reach levels way beyond (up to 10^{12}/cm^2/yr).

1.3   Particle Interaction Mechanisms

This section serves as an introduction to particle-matter interactions, focused mainly on important mechanisms capable of inducing radiation effects on electronic components. These effects can be broadly classified into two groups: cumulative effects, which encompass total ionizing dose (TID) and displacement damage (DD) and single-event effects (SEEs), a category of effects where a single particle impact can disrupt the proper operation of electronic devices. Regardless of the specific effect on the electronic device, it always originates from one or more particles that deposit energy in the device. In this book, our focus is primarily on the study and mitigation of single-event effects, more specifically on the nondestructive effects known as *soft errors*. Therefore, the subsequent sections will be dedicated to comprehending these effects in greater detail.

Numerous types of particles have the capability to deposit energy in a device and induce SEEs. Radiation-matter interactions are strongly dependent on the characteristics of the particle, including its charge, mass, and energy. For instance, charged particles like protons engage in both nuclear reactions and Coulombic interactions with the orbital electrons of the target nucleus. Conversely, neutrons, being uncharged, can traverse the electronic clouds of an atom without interaction, making them highly penetrating particles that solely interact and lose energy with atomic nuclei. Figure 1.8 outlines the main Coulombic and nuclear interactions, as well as photon-matter interactions, which are essential for understanding the impact of radiation on electronics.

- **Coulombic Interactions:** *Ionization*, when a charged particle (e.g., proton) interacts with the electric field of an orbital electron, it can remove it from its orbit, creating an electron-hole pair, where the vacancy left behind by the ejected electron acts like a positive charge carrier (hole); *Excitation*, occurs when the interaction with the electric field only elevates the electron to a higher energy level within the same atom and upon returning to its ground state, the electron releases energy in the form of a photon; *Elastic collision*, when a charged particle is deflected by the atomic nucleus and the total kinetic energy is conserved; *Bremsstrahlung*, occurs when a charged particle accelerates or decelerates, emitting electromagnetic radiation (photons) due to this change in motion.
- **Nuclear Reactions:** *Neutron absorption or capture*, occurs when a low-energy neutron collides with a nucleus, leading to the emission of gamma rays as the excess energy is released; in *elastic scattering*, the incident particle collides with the nucleus, transferring some kinetic energy to the nucleus. The nucleus recoils, while the incident particle is scattered in a different direction; however, the total kinetic energy of the system remains conserved; on the other hand, in *inelastic scattering*, the total kinetic energy is not conserved and the collision can induce the emission of photons or ejection of electrons; a similar phenomenon is observed in *non-elastic scattering*, where the nucleus is fragmented into secondary particles, neutrons, or other ions species.
- **Photon-matter interactions:** *Photoelectric effect*, a photon transfers all its energy to a tightly bound orbital electron, ejecting it from the atom with a kinetic energy equal to the difference between the energy of the incident photon and the binding energy; *compton scattering* occurs when a photon interacts with a relatively free electron (compared to tightly bound inner-shell electrons), imparting a portion of its energy to the electron, which recoils with some kinetic

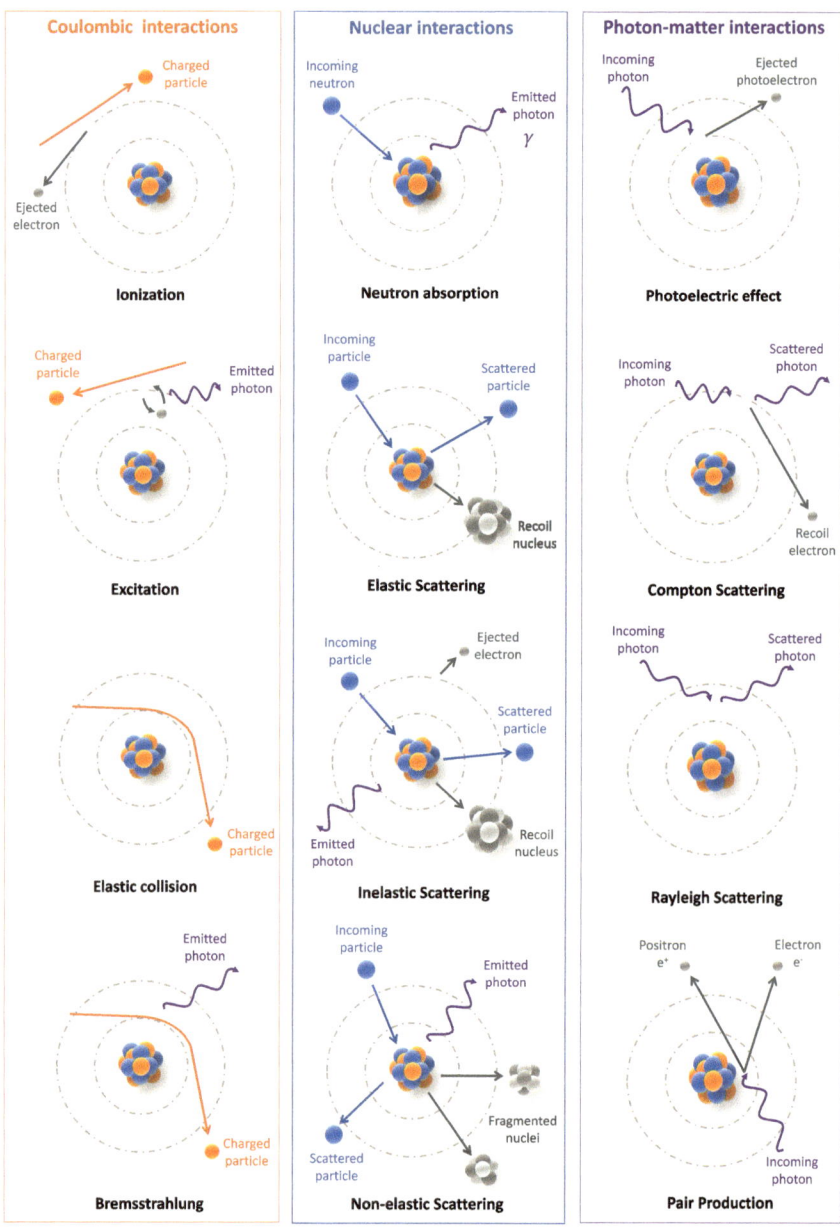

Fig. 1.8 Simplified representation of main particle-matter mechanisms (Coulombic, nuclear, and photon-matter interactions)

energy. The remaining energy of the photon is carried away by a scattered photon with a longer wavelength (lower energy) compared to the incident photon; *Rayleigh scattering*, also known as coherent scattering, occurs when a low-energy photon interacts with an atom and is scattered elastically. The scattered photon retains its energy but deflects in a different direction; lastly, *pair production*, high-energy photons (typically exceeding 1.022 MeV) can interact with the strong electric field near a nucleus, leading to the creation of an electron-positron pair. The positron is the antiparticle of the electron and has the same mass but a positive charge. The minimum energy requirement of at least 1.022 MeV for a photon arises from the fact that the resting mass of both the electron and positron, when expressed in energy units, is 0.511 MeV each. Consequently, for the creation of an electron-positron pair, a minimum total energy of 1.022 MeV is necessary, corresponding to twice the energy of a single electron or positron.

1.4 Energy Deposition

Originally, single-event effects (SEEs) were primarily associated with neutrons, protons, or heavier ions. It is worth noting, however, that certain studies have shown that SEEs can also be triggered by electrons [33–36] or muons [14–17]. Nevertheless, for the purposes of our discussion, these particles will not be considered. SEEs primarily result from ionization, occurring directly with ions and protons, and indirectly in the case of neutrons due to nuclear reactions. While ions and protons can directly ionize the target material, they can also engage in nuclear reactions with the atomic nucleus. For heavy ions, the contribution from nuclear reactions is typically insignificant compared to direct ionization. However, for protons, both direct and indirect ionization processes are significant. Figure 1.9 illustrates the direct and indirect ionization processes for a heavy ion and a neutron, respectively.

Historically, the direct ionization of protons was not a major concern as they had limited capability to ionize matter directly. Consequently, only high-energy protons ($E > 20\,\text{MeV}$) were considered a threat to electronic systems due to their capacity to induce nuclear reactions in silicon and produce secondary ions with higher ionization potential. However, in deeply scaled technologies, low-energy protons ($E < 3\,\text{MeV}$) have been identified as significant contributors to soft error rates in space applications and beyond [37–39].

The loss of energy is generally quantified by the concept of *stopping power*, S, also called *specific energy loss*, which is defined as the differential energy loss dE dividing by the differential path length dx:

$$S = -\frac{\text{dE}}{\text{dx}} \tag{1.1}$$

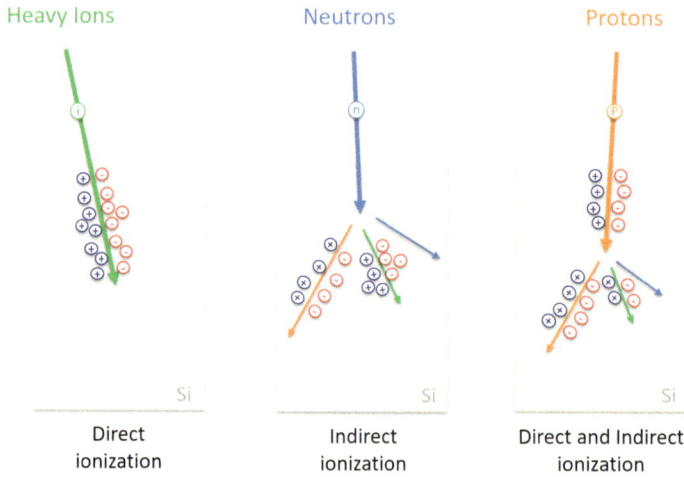

Fig. 1.9 Representation of direct ionization and indirect ionization for heavy ions, neutrons, and protons interaction with matter (Si stands for silicon)

where the minus sign is a common way to deal with a positive quantity (the energy loss dE is a negative variation of energy). The stopping power unit is given in megaelectronvolts per centimeter (MeV/cm). In the radiation effect community, the stopping power is generally divided by the mass density of the material, making the new quantity independent of its phase. This new quantity, called *mass stopping power*, S_m is then:

$$S_m = -\frac{1}{\rho}\frac{dE}{dx} \tag{1.2}$$

where ρ is the density of the target material. For electronics, the silicon density is used, $\rho_{Si} = 2.32 \text{ g} \cdot \text{cm}^{-3}$. The mass stopping power unit is given in megaelectronvolts square centimeter per milligram (MeV $\cdot$ cm^2/mg).

Another important quantity for charged particle is the *linear energy transfer*, or **LET**. While the stopping power measures the energy loss of the particle, the LET measures the energy absorbed in matter. The difference comes from the energy loss by radiation, which is actually not significant for ions. Through misuse of language, the word *LET* is often use instead of *stopping power* or even instead of *mass stopping power*.

The stopping power depends not only on the particle type but also on its energy and the target material where ionization takes place. The variation of mass stopping power as a function of the ion energy in silicon was calculated for different ions using the stopping and range of ions in matter (SRIM) code [40] as shown in Fig. 1.10.

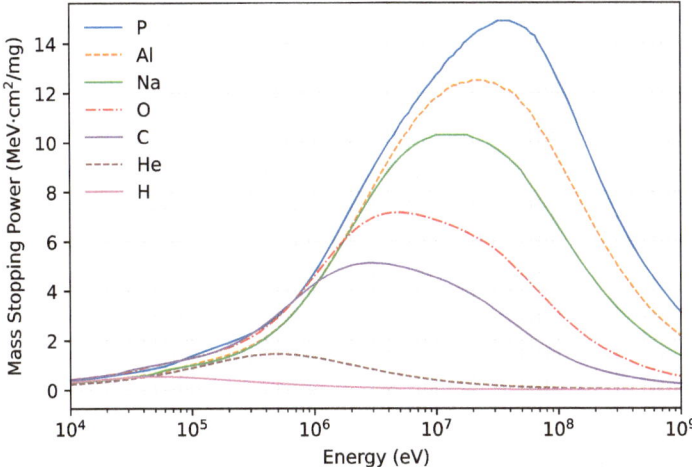

Fig. 1.10 Variation of the stopping power as a function of energy for different ions in silicon (Data obtained from SRIM [40])

The stopping power of a particle increases with energy until it reaches the *Bragg peak*, which represents the maximum value of stopping power. It is important to note that the Bragg peak for hydrogen (protons) is significantly lower compared to heavier ions like phosphorus. In general, the stopping power at the Bragg peak is higher for ions with a higher atomic number (Z). This is because the energy loss is primarily governed by the Coulomb effect, and at higher Z, the interaction between the particle and the medium is stronger. For energies above the Bragg peak, the stopping power decreases as the particle energy increases. It is interesting to observe that different ion species with different energies can yield the same stopping power value. This is because energy loss involves interactions with both electrons and nuclei of the medium. Therefore, the stopping power is typically considered as the sum of two contributions:

$$S_m = -\frac{1}{\rho}\frac{dE}{dx}\Big|_{elec} - \frac{1}{\rho}\frac{dE}{dx}\Big|_{nuc} \tag{1.3}$$

The first term represents the interaction of the particle with electrons. When sufficient energy is transferred, it leads to the generation of electron-hole pairs. This process is known as *ionization*, as an electron moves from the valence band to the conduction band. The second term accounts for the interaction of the particle with the nuclei of the medium. This interaction can be mediated by either the Coulomb force or the strong nuclear force. In the case of Coulomb interactions, the target nucleus can undergo vibration or recoil, leading to *displacement damage*. Alternatively, if the interaction is governed by the strong force, a *nuclear reaction* can occur, causing the nucleus to break up into fragments.

While neutrons, protons, and heavy ions interact with matter in distinct ways, they all ultimately result in the ionization of the material, either directly or indirectly. This ionization process can potentially trigger SEEs in electronic devices. Therefore, understanding the stopping power of these ions, both primary and secondary, is of prime importance for the assessment of their impact and the mitigation of the associated risks.

Another crucial quantity of interest for charged particles is the *range*, denoted as R. It is defined as the distance the particle can travel before being at rest. The range is related to the stopping power through the following relationship:

$$R = \int_{E_{init}}^{0} \frac{1}{-\frac{dE}{dx}} \, dx \tag{1.4}$$

where E_{init} is the initial energy of the particle. Particles with high stopping power experience significant energy loss, resulting in a short range of travel. This can be observed in Fig. 1.11, which presents examples of particle ranges calculated using SRIM. The figure illustrates how the range varies for different particles, highlighting the impact of their stopping power on the distance they can travel in a material.

The energy deposition in the semiconductor leads to a nearly linear path of *electron-hole pairs* (ehp). The minimum energy required to generate an electron-

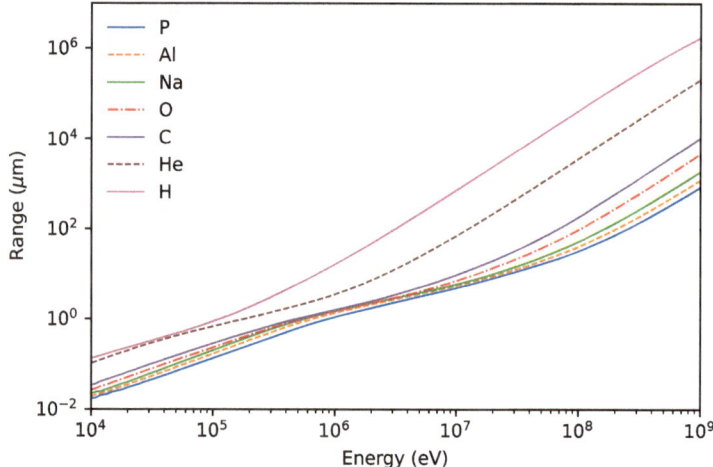

Fig. 1.11 Variation of the range as a function of energy for different ions in silicon (Data obtained from SRIM [40])

Table 1.2 Bandgap energy E_g and ionization energy E_{ehp} for different semiconductor materials

Material	Symbol	Bandgap energy E_g (eV)	Ionization energy E_{ehp}(eV)
Germanium	Ge	0.66	2.4
Silicon	Si	1.11	3.6
Indium phosphide	InP	1.34	4.2
Gallium arsenide	GaAs	1.43	4.5
Gallium phosphide	GaP	2.26	6.7
Silicon carbide	3C-SiC	2.35	7.0
	6H-SiC	3.08	9.0
	4H-SiC	3.28	9.5
Gallium nitride	GaN	3.4	9.8
Aluminum nitride	AlN	6.24	17.6

hole pair can be estimated based on the *bandgap energy E_g* of the material using Eq. 1.5 [41]:

$$E_{ehp} = 2.73 E_g + 0.55 \qquad (1.5)$$

For silicon-based devices, the silicon bandgap energy E_g is equal to 1.11 eV; therefore, based on Eq. 1.5, the average *ionization energy E_{ehp}* in silicon is approximately 3.6 eV/ehp. In Table 1.2, for different semiconductor materials, the bandgap energy E_g and the corresponding ionization energy E_{ehp} calculated using Eq. 1.5 are presented. The higher the bandgap energy of the material, the higher is the energy required to ionize the matter and create electron-hole pairs. This explains why wide-bandgap electronics such as the SiC- and GaN-based devices show a better radiation performance overall.

Based on the assumption that the stopping power is constant as a consequence of the small ionization path length l within the sensitive volume of a device, the deposited energy can be converted in *deposited charge Q_D* by following Eq. 1.6:

$$Q_D = \frac{q \cdot S \cdot \rho \cdot l}{E_{ehp}} \qquad (1.6)$$

where q is the elementary charge and S is the so-called *surface mass stopping power* which refers to the mass stopping power value when the particle enters the sensitive volume, i.e., after traveling and losing energy through all the back-end-of-line (BEOL) layers. The deposited charge Q_D needs to be collected by a sensitive node of a circuit and exceeds the minimum charge necessary to induce a SEE, known as *critical charge Q_{crit}*. The charge collection mechanisms are briefly described in the following section.

Another quite relevant aspect, when studying radiation effects on electronics, is the temperature dependence of many mechanisms at the device and circuit levels. For instance, the bandgap energy of a semiconductor material is temperature dependent and can be described using Varshni's Equation [42]:

$$E_g[T] = E_g[0] - \frac{\alpha \cdot T^2}{T + \beta} \tag{1.7}$$

where $E_g[0]$ is the bandgap energy for temperature T equals to 0 K, while α and β are fitting parameter to experimental data. These three parameters are presented in Table 1.3 for the three main semiconductor materials. In Fig. 1.12, the bandgap and ionization energy for silicon and gallium arsenide (GaAs) are shown for a wide range of temperature measure in K and calculated with Eq. 1.7. As mentioned previously, lower ionization energy is observed in Si due to its lower bandgap energy when compared to GaAs. However, as the temperature is increased, both materials show a reduction in the ionization energy. Although the decrease is minimal, it can increase the number of electron-hole pairs that a particle can generate and increase the total deposited charge. Other mechanisms are also affected by temperature, as it will be shown in the following chapters, such as carrier mobility and resistance.

Table 1.3 Fitting parameters for Varshni's equation to consider the temperature dependence of bandgap energy of Ge, Si, GaAs, and 6H-SiC

Material	Symbol	$E_g[0]$	α	β
Germanium	Ge	0.7412	4.561	210
Silicon	Si	1.557	7.021	1108
Gallium arsenide	GaAs	1.5216	8.871	572
Silicon carbide	6H-SiC	3.024	−0.3055	−311

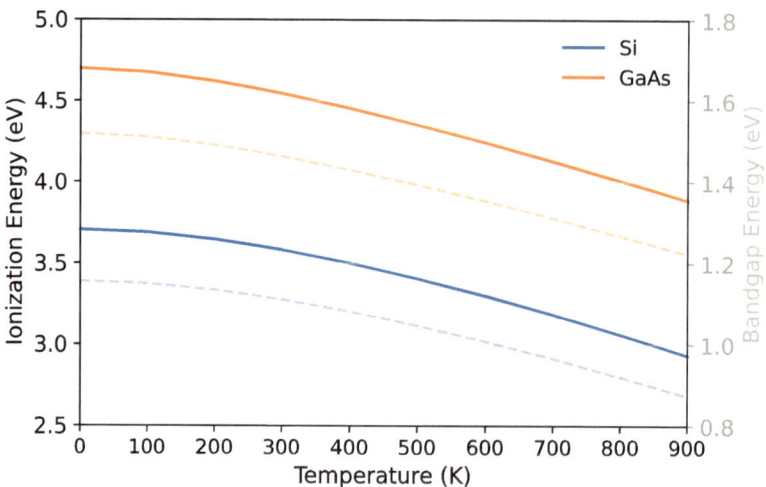

Fig. 1.12 Ionization energy (solid lines) and bandgap energy (dashed lines) for silicon and gallium arsenide as a function of temperature

Besides depending on the material, the *ionization energy* E_{ehp} is also **temperature dependent**. As the temperature increases, the bandgap energy reduces and, therefore, the ionization energy is also reduced.

1.5 Charge Collection

After the energy deposition in the semiconductor material, the generated carriers are transported and collected by the junctions of the device. In the context of SEE, there are two main transport mechanisms that play a crucial role: the *drift* and the *diffusion*. Figure 1.13 illustrates the ionization process of an ion in a reverse-biased p-n junction and the subsequent carrier transport mechanisms.

Drift is a transport mechanism that is primarily governed by the electric field existing within the p-n junctions of sensitive electronic devices. When a particle directly impacts the sensitive collecting area of the circuit, the resulting carriers experience rapid collection due to the influence of the high electric field present in the reverse-biased p-n junction. In semiconductor materials, the drift current J_{drift} can be described as a function of the applied electrical field $\mathscr{E}$ and the conductance G, as expressed by the equation:

$$J_{drift} = G \cdot \mathscr{E} \tag{1.8}$$

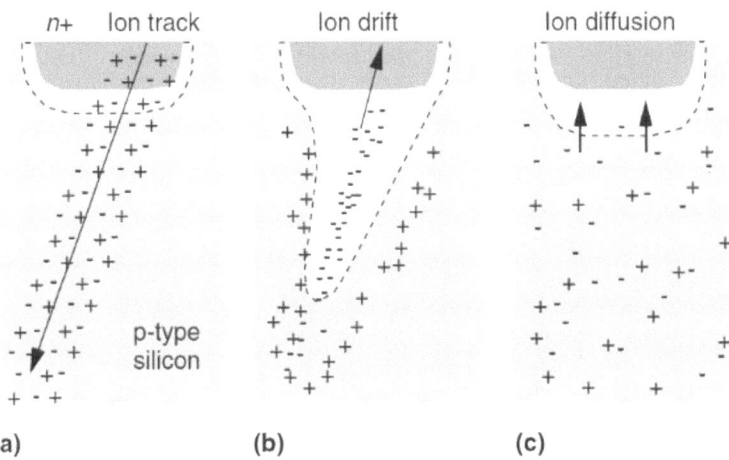

Fig. 1.13 Charge collection mechanisms due to an ion strike in a p-n junction (Adapted from [43])

Finally, by knowing that the conductance G is proportional to the the carrier mobility μ and to the carrier concentrations n and p, the total drift current J_{drift} can be described as [44]:

$$J_{drift} = q(\mu_n n + \mu_p p) \cdot \mathscr{E} \tag{1.9}$$

where q is the elemental charge ($q = 1.6 \times 10^{-19}C$).

On the other hand, diffusion is a carrier transport mechanism governed by the carrier concentration gradients. Therefore, whenever there is a gradient of carriers, they are transported from regions with high to low concentrations to reach a state of uniformity [44]. The diffusion current J can be obtained based on Fick's Law:

$$J_{Dn} = q D_n \frac{dn}{dx} \tag{1.10}$$

$$J_{Dp} = -q D_p \frac{dp}{dx} \tag{1.11}$$

where D_n and D_p are the diffusion coefficients and dn/dx and dp/dx are gradient of carrier concentrations for electrons and holes, respectively. As the diffusion is governed by random thermal motion of carriers and scattering, it has a direct relationship with the mobility of carriers μ and, therefore, the diffusion coefficients can be described as [44]:

$$D_{n,p} = \left(\frac{kT}{q}\right) \mu_{n,p} \tag{1.12}$$

where, at room temperature (300 K), (kT/q) is equal to 0.0259 V.

Considering the one-dimensional case, the total current J can be calculated using Eq. 1.9 for the drift component, and for the diffusion component using Eqs. 1.10 and 1.14 as shown below:

$$J_n = q \cdot \left(\mu_n n \mathscr{E} + D_n \frac{dn}{dx}\right) \tag{1.13}$$

$$J_p = q \cdot \left(\mu_p p \mathscr{E} - D_p \frac{dp}{dx}\right) \tag{1.14}$$

If we want to determine the evolution of carriers concentrations during time, we need to solve the transport equations that are given by:

$$q \frac{\partial n}{\partial t} = \nabla J_n - q \times R \tag{1.15}$$

$$q \frac{\partial p}{\partial t} = -\nabla J_p + q \times R \tag{1.16}$$

where R is the recombination rate of electron-hole pairs.

Solving transport equations is a long and complex process that requires
to perform approximations and use numerical tools, called technology
computer-aided design (TCAD) tools.

Another very interesting effect occurs when the particle strikes in or near the p-n
junction, as shown in Fig. 1.13b. The depletion region, also known as the space
charge region, is increased, and therefore, the drift collection efficiency is also
increased. This phenomenon is known as *funneling effect* and together with drift,
they are two very fast processes due to the presence of the high electric field in
the junction. Following these processes, the remaining carriers are collected by
diffusion process or they go through recombination.

Understanding these intricate processes is crucial for comprehending the behavior of semiconductor devices, particularly in relation to their response to particle
interactions. By examining the interplay between drift, funneling effect, diffusion,
and recombination, we can gain valuable insights into the overall performance and
reliability of these devices.

1.5.1   Charge Sharing and Pulse Quenching Effect

As technology advances and transistors are placed in closer proximity to each
other, the critical charge required to trigger an SEE is reduced. Consequently, a
single incident particle is able to induce sufficient charge collection from multiple
neighboring electrodes. This phenomenon is known as *charge sharing effect*. The
work developed by Amusan et al. [45] provides an analysis of the charge sharing
effect in adjacent N-channel metal-oxide semiconductor (NMOS) and P-channel
metal-oxide semiconductor (PMOS) transistors based on TCAD simulations. The
130 nm CMOS devices, based on the International Business Machines corporation
(IBM) twin well technology, were characterized using advanced simulation tools
such as Synopsis DEVISE and DESSIS. In the simulation setup, a particle hit was
precisely targeted at the center of the drain region, perpendicular to the surface of
the device structure. It is worth noting that for the purpose of these simulations,
the influence of angular effects was omitted in the analysis. The authors used two
particular notations: **active device**, which refers to the device directly struck by the
particle and actively collecting the carriers, and the **passive device**, which represents
the device not directly impacted by the particle but passively collecting the diffused
carriers. In Fig. 1.14, the collected charge is shown for the active and passive devices
when PMOS and NMOS transistors are used.

The results of this study clearly demonstrate that passive PMOS devices are
capable of collecting a greater amount of charge compared to passive NMOS
devices. Specifically, the passive PMOS device collected approximately 40% of the
charge collected by the active PMOS device, while the passive NMOS transistor

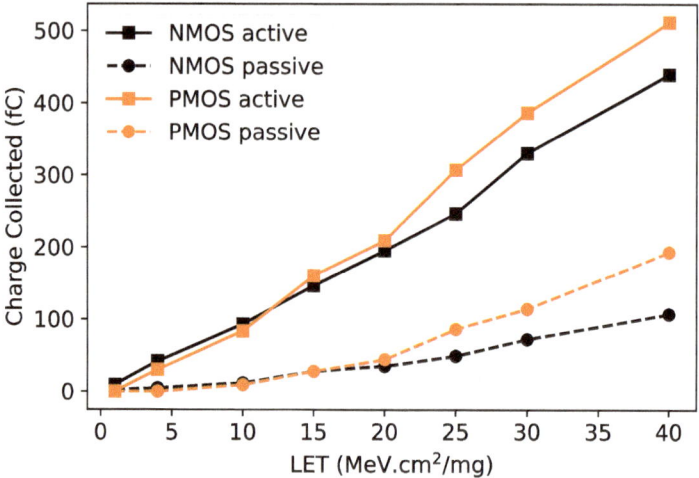

Fig. 1.14 Charge collected by PMOS and NMOS for normal particle hits at the center of the drain region of the active device (Data from [45])

collected less than 25% of the charge. The authors attribute this discrepancy to the disparity in carrier diffusion coefficients between electrons and holes, as well as the bipolar amplification effect that enhances charge collection in PMOS devices [45, 46].

While the charge sharing mechanism is responsible for the increased SEE sensitivity in deeply scaled technologies, it has also been shown to reduce the pulse width of SETs in combinational cells [47, 48]. Due to the similar time constant for the circuit delay and the diffusion process, the radiation-induced transient is able to activate the charge collection in electrodes from following stage of circuits in such a way that the resultant transient is shortened (i.e., quenched). This phenomenon is referred to as the pulse quenching effect (PQE). To observe the PQE in a circuit, an inverting relationship between logic stages is required. Both charge sharing and the pulse quenching effect are intricate and critical mechanisms that heavily depend on the specific device technology and design strategies employed. Therefore, acquiring a comprehensive understanding of their impact is of utmost importance for accurately assessing the sensitivity of modern electronic components to single-event effects.

1.6 Summary

In this chapter, we explored the fundamental concepts surrounding the study of radiation effects in electronics. Our exploration begins by unraveling the complex dynamics and composition of the space and atmospheric radiation environment. At the heart of this cosmic interplay is the Sun, which acts as the primary modulator of solar particle radiation and galactic cosmic rays. For missions focused on Earth's vicinity, it becomes crucial to consider the impact of the Van Allen radiation belts on on-board electronic systems. Furthermore, in low-orbit and atmospheric applications like satellites and aviation, the South Atlantic Anomaly region exhibits a heightened proton flux, necessitating careful attention.

Depending on the type of incident radiation, direct and indirect ionization processes can occur within electronic components, resulting in energy deposition and charge collection by critical device electrodes. In advanced technology nodes, a single particle strike has the potential to impact multiple devices within a chip, resulting in the diffusion of charges across multiple critical nodes—a phenomenon known as the charge sharing effect. Looking ahead, the subsequent chapter will delve into the radiation-induced effects at the circuit level, specifically focusing on the well-known single-event effects (SEEs).

Highlights

- Three main natural sources of radiation should be considered in space and atmospheric applications: the *solar energetic particles* (SEPs), *galactic cosmic rays* (GCRs), and geomagnetically trapped particle radiation (*Van Allen's radiation belts*).
- The Sun's activity is the main radiation modulator of the radiation environment near Earth, but also impacts ground-level applications, such as the Carrington event.
- A misalignment between the Earth's geomagnetic and rotational axes generates a weakness in the geomagnetic field over South America, leading to higher particle fluxes in this region known as the South Atlantic Anomaly (SAA).
- With the technology scaling, neutrons but also muons are considered a threat to the electronic system reliability.
- In an accelerator environment, there are also three main sources of radiation: beam interaction with residual gas, beam interaction with machine elements, and debris from collisions in physics experiments.
- The high-energy hadrons (HEH) and neutrons are the most abundant and relevant particles in the LHC environment in terms of inducing electronic failures.
- An energetic particle can deposit energy either through direct ionization (ions, protons, etc.) or by indirect ionization (protons and neutrons).
- Two main mechanisms are responsible for charge collection in a device: drift due to the electric field present in the p-n junctions and diffusion due to the carrier concentration gradients.
- With the miniaturization of transistor technology, a single particle is able to deposit energy in multiple devices, and therefore charge sharing effect is prominent in highly scaled technologies.

References

1. NASA website. https://www.nasa.gov/
2. ECSS-E ESA. St-10-04c rev. 1 space engineering. *Space environment. ESA-ESTEC, Noordwijk, Jun*, 2020.
3. MA Shea and DF Smart. A comparison of energetic solar proton events during the declining phase of four solar cycles (cycles 19–22). *Advances in Space Research*, 16 (9): 37–46, 1995.
4. Space weather prediction center. https://www.swpc.noaa.gov/
5. Scott E Forbush. World-wide cosmic ray variations, 1937–1952. *Journal of Geophysical Research*, 59 (4): 525–542, 1954.
6. IV Dorman and LI Dorman. Solar wind properties obtained from the study of the 11-year cosmic-ray cycle: 1. *Journal of Geophysical Research*, 72 (5): 1513–1520, 1967.
7. K Nagashima and I Morishita. Twenty-two year modulation of cosmic rays associated with polarity reversal of polar magnetic field of the sun. *Planetary and Space Science*, 28 (2): 195–205, 1980.
8. Richard C Carrington. Description of a singular appearance seen in the sun on September 1, 1859. *Monthly Notices of the Royal Astronomical Society*, 20: 13–15, 1859.
9. Edward W Cliver and William F Dietrich. The 1859 space weather event revisited: limits of extreme activity. *Journal of Space Weather and Space Climate*, 3: A31, 2013.
10. Stuart Clark. Solar surprise. *New Scientist*, 254 (3387): 38–41, 2022.
11. Ryuho Kataoka, Daikou Shiota, Hitoshi Fujiwara, Hidekatsu Jin, Chihiro Tao, Hiroyuki Shinagawa, and Yasunobu Miyoshi. Unexpected space weather causing the reentry of 38 starlink satellites in february 2022. doi: https://doi.org/10.31223/X5GH0X.
12. Marco Durante and Francis A Cucinotta. Physical basis of radiation protection in space travel. *Reviews of Modern Physics*, 83 (4): 1245, 2011.
13. The omere 5.3 software by trad and cnes. URL http://www.trad.fr/en/space/omere-software
14. Brian D Sierawski, Marcus H Mendenhall, Robert A Reed, Michael A Clemens, Robert A Weller, Ronald D Schrimpf, Ewart W Blackmore, Michael Trinczek, Bassam Hitti, Jonathan A Pellish, et al. Muon-induced single event upsets in deep-submicron technology. *IEEE Transactions on Nuclear Science*, 57 (6): 3273–3278, 2010.
15. G Hubert, L Artola, and D Regis. Impact of scaling on the soft error sensitivity of bulk, FDSOI and FinFET technologies due to atmospheric radiation. *Integration, the VLSI journal*, 50: 39–47, 2015.
16. Angelo Infantino, Ewart W Blackmore, Markus Brugger, Rubén García Alía, Matthew Stukel, and Michael Trinczek. Fluka Monte Carlo assessment of the terrestrial muon flux at low energies and comparison against experimental measurements. *Nuclear Instruments and Methods in Physics Research Section A: Accelerators, Spectrometers, Detectors and Associated Equipment*, 838: 109–116, 2016.
17. Takashi Kato, Motonobu Tampo, Soshi Takeshita, Hiroki Tanaka, Hideya Matsuyama, Masanori Hashimoto, and Yasuhiro Miyake. Muon-induced single-event upsets in 20-nm SRAMs: Comparative characterization with neutrons and alpha particles. *IEEE Transactions on Nuclear Science*, 68 (7): 1436–1444, 2021.
18. Peter KF Grieder. *Cosmic rays at Earth*. Elsevier, 2001.
19. Janet L Barth, CS Dyer, and EG Stassinopoulos. Space, atmospheric, and terrestrial radiation environments. *IEEE Transactions on nuclear science*, 50 (3): 466–482, 2003.
20. MS Gordon, P Goldhagen, KP Rodbell, TH Zabel, HHK Tang, JM Clem, and Paul Bailey. Measurement of the flux and energy spectrum of cosmic-ray induced neutrons on the ground. *IEEE Transactions on Nuclear Science*, 51 (6): 3427–3434, 2004.
21. Matteo Cecchetto. *Experimental and simulation study of neutron-induced single event effects in accelerator environment and implications on qualification approach*. PhD thesis, Montpellier, 2021.

22. Oliver Sim Brüning, Paul Collier, P Lebrun, Stephen Myers, Ranko Ostojic, John Poole, and Paul Proudlock. *LHC Design Report*. CERN Yellow Reports: Monographs. CERN, Geneva, 2004. doi: https://doi.org/10.5170/CERN-2004-003-V-1. URL https://cds.cern.ch/record/782076

23. Rubén García Alía, Markus Brugger, Francesco Cerutti, Salvatore Danzeca, Alfredo Ferrari, Simone Gilardoni, Yacine Kadi, Maria Kastriotou, Anton Lechner, Corinna Martinella, et al. LHC and HL-LHC: Present and future radiation environment in the high-luminosity collision points and rha implications. *IEEE Transactions on Nuclear Science*, 65 (1): 448–456, 2017.

24. Kacper Biłko, Cristina Bahamonde Castro, Markus Brugger, Rubén García Alía, Yacine Kadi, Anton Lechner, Giuseppe Lerner, and Oliver Stein. Radiation environment in the LHC Arc sections during run 2 and future HL-LHC operations. *IEEE Transactions on Nuclear Science*, 67 (7): 1682–1690, 2020.

25. Oliver Stein, Kacper Bilko, Markus Brugger, Salvatore Danzeca, Diego Di Francesca, R Garcia Alia, Yacine Kadi, G Li Vecchi, Corinna Martinella, et al. A systematic analysis of the prompt dose distribution at the large hadron collider. In *9th Int. Particle Accelerator Conf. (IPAC'18), Vancouver, BC, Canada, April 29-May 4, 2018*, pages 2036–2038. JACOW Publishing, Geneva, Switzerland, 2018.

26. Andrea Coronetti, Rubén García Alía, Jan Budroweit, Tomasz Rajkowski, Israel Da Costa Lopes, Kimmo Niskanen, Daniel Söderström, Carlo Cazzaniga, Rudy Ferraro, Salvatore Danzeca, et al. Radiation hardness assurance through system-level testing: Risk acceptance, facility requirements, test methodology, and data exploitation. *IEEE Transactions on Nuclear Science*, 68 (5): 958–969, 2021.

27. Ygor Aguiar, Andrea Apollonio, Giuseppe Lerner, Francesco Cerutti, Marta Sabaté-Gilarte, Salvatore Danzeca, Daniel Prelipcean, and Ruben García Alía. Radiation to electronics impact on CERN LHC operation: Run 2 overview and HL-LHC outlook. 2021.

28. Ygor Aguiar, Andrea Apollonio, Giuseppe Lerner, Matteo Cecchetto, Jean-Baptiste Potoine, Matteo Brucoli, Salvatore Danzeca, Ruben García Alía, Kacper Biłko, Alessandro Zimmaro, et al. Implications and mitigation of radiation effects on the CERN SPS operation during 2021. *JACoW IPAC*, 2022: 740–743, 2022.

29. FLUKA website. https://fluka.cern

30. G. Battistoni et al. Overview of the FLUKA code. *Annals Nucl. Energy*, 82: 10–18, 2015. doi: https://doi.org/10.1016/j.anucene.2014.11.007.

31. T. Böhlen et al. The FLUKA Code: Developments and Challenges for High Energy and Medical Applications. *Nuclear Data Sheets*, 120: 211–214, 06 2014. doi: https://doi.org/10.1016/j.nds.2014.07.049.

32. R Garcia Alia et al. Radiation fields in high energy accelerators and their impact on single event effects. *Diss. Montpellier U*, 2014.

33. MP King, RA Reed, RA Weller, MH Mendenhall, RD Schrimpf, BD Sierawski, AL Sternberg, B Narasimham, JK Wang, E Pitta, et al. Electron-induced single-event upsets in static random access memory. *IEEE Transactions on Nuclear Science*, 60 (6): 4122–4129, 2013.

34. Maris Tali, Rubén García Alía, Markus Brugger, Veronique Ferlet-Cavrois, Roberto Corsini, Wilfrid Farabolini, Ali Mohammadzadeh, Giovanni Santin, and Ari Virtanen. High-energy electron-induced SEUs and Jovian environment impact. *IEEE Transactions on Nuclear Science*, 64 (8): 2016–2022, 2017.

35. Pablo Caron, Christophe Inguimbert, Laurent Artola, Nathalie Chatry, Nicolas Sukhaseum, Robert Ecoffet, and Françoise Bezerra. Physical mechanisms inducing electron single-event upset. *IEEE Transactions on Nuclear Science*, 65 (8): 1759–1767, 2018.

36. Pablo Caron, Christophe Inguimbert, Laurent Artola, Françoise Bezerra, and Robert Ecoffet. Role of electron-induced coulomb interactions to the total SEU rate during earth and juice missions. *IEEE Transactions on Nuclear Science*, 68 (8): 1607–1612, 2021.

37. Kenneth P Rodbell, David F Heidel, Henry HK Tang, Michael S Gordon, Phil Oldiges, and Conal E Murray. Low-energy proton-induced single-event-upsets in 65 nm node, silicon-on-insulator, latches and memory cells. *IEEE Transactions on Nuclear Science*, 54 (6): 2474–2479, 2007.

38. Rubén García Alía, Maris Tali, Markus Brugger, Matteo Cecchetto, Francesco Cerutti, Andrea Cononetti, Salvatore Danzeca, Luigi Esposito, Pablo Fernández-Martínez, Simone Gilardoni, et al. Direct ionization impact on accelerator mixed-field soft-error rate. *IEEE Transactions on Nuclear Science*, 67 (1): 345–352, 2019.

39. Andrea Coronetti, Rubén Garcìa Alìa, Jialei Wang, Maris Tali, Matteo Cecchetto, Carlo Cazzaniga, Arto Javanainen, Frédéric Saigné, and Paul Leroux. Assessment of proton direct ionization for the radiation hardness assurance of deep submicron srams used in space applications. *IEEE Transactions on Nuclear Science*, 68 (5): 937–948, 2021.

40. James F Ziegler, Matthias D Ziegler, and Jochen P Biersack. Srim–the stopping and range of ions in matter (2010). *Nuclear Instruments and Methods in Physics Research Section B: Beam Interactions with Materials and Atoms*, 268 (11–12): 1818–1823, 2010.

41. RC Alig and S Bloom. Electron-hole-pair creation energies in semiconductors. *Physical review letters*, 35 (22): 1522, 1975.

42. Yatendra Pal Varshni. Temperature dependence of the energy gap in semiconductors. *Physica*, 34 (1): 149–154, 1967.

43. Robert C Baumann. Radiation-induced soft errors in advanced semiconductor technologies. *Device and Materials Reliability, IEEE Transactions on*, 5 (3): 305–316, 2005.

44. SM Sze and Kwok K Ng. Physics of semiconductor devices, John Wiley & Sons. *New York*, 68, 1981.

45. O. A. Amusan, A. F. Witulski, L. W. Massengill, B. L. Bhuva, P. R. Fleming, M. L. Alles, A. L. Sternberg, J. D. Black, and R. D. Schrimpf. Charge collection and charge sharing in a 130 nm cmos technology. *IEEE Transactions on Nuclear Science*, 53 (6): 3253–3258, Dec 2006. ISSN 0018-9499. doi: https://doi.org/10.1109/TNS.2006.884788.

46. B. Liu, S. Chen, B. Liang, Z. Liu, and Z. Zhao. Temperature dependency of charge sharing and mbu sensitivity in 130-nm cmos technology. *IEEE Transactions on Nuclear Science*, 56 (4): 2473–2479, Aug 2009. ISSN 0018-9499. doi: https://doi.org/10.1109/TNS.2009.2022267.

47. J. R. Ahlbin, L. W. Massengill, B. L. Bhuva, B. Narasimham, M. J. Gadlage, and P. H. Eaton. Single-event transient pulse quenching in advanced cmos logic circuits. *IEEE Transactions on Nuclear Science*, 56 (6): 3050–3056, Dec 2009. ISSN 0018-9499. doi: https://doi.org/10.1109/TNS.2009.2033689.

48. N. M. Atkinson, A. F. Witulski, W. T. Holman, J. R. Ahlbin, B. L. Bhuva, and L. W. Massengill. Layout technique for single-event transient mitigation via pulse quenching. *IEEE Transactions on Nuclear Science*, 58 (3): 885–890, June 2011. ISSN 0018-9499. doi: https://doi.org/10.1109/TNS.2010.2097278.

Chapter 2
Introduction to Single-Event Effects

2.1 Context and Overview

As presented in the previous chapter, the reliability of electronic circuits is subject to physical damage or functional failures due to the influence of the environment, such as the presence of atmospheric or space radiation [1]. The energy deposition of a single energetic particle in the sensitive areas of a circuit can lead to destructive or nondestructive mechanisms, known as single-event effects. Initially, the first studies on circuit reliability under the stress of radiation effects were considered a primary concern of extreme relevance only in projects developed for military or space applications due to their harsh environments. Back in 1962, the work developed in [2] was the first study to predict that galactic cosmic radiation could become a threat to circuit design as the technology is scaled down into the nanometer world. And, only later in 1975, Binder et al. [3] were able to identify anomalies in the bit storage in flip-flop circuits used in a satellite system and attributed to the cosmic radiation effects.

Besides the radiation effects observed in space applications, these irregularities in the circuit operation were also identified at sea level as early as 1978 [4]. However, the root cause for these anomalies that were observed in memory circuits was associated to the alpha particles emitted from the uranium and thorium composition which were naturally present in the package material surrounding the devices. This chapter used for the very first time the term *soft errors* to associate the nondestructive radiation effects in electronics, and it is still largely adopted in the research community. In the following year, Guenzer et al. [5] have shown that neutrons and protons can also induce upsets in memory elements when they trigger nuclear reactions within the circuit material. It was in this chapter that the term *single-event upset* (SEU) was first adopted to address the bit flips observed in memory circuits, and it has been largely used since then. In the next section, the fundamental SEU mechanisms necessary to understand and investigate their effects on current technologies will be explained in detail.

© The Author(s) 2025
Y. Quadros de Aguiar et al., *Single-Event Effects, from Space to Accelerator Environments*, https://doi.org/10.1007/978-3-031-71723-9_2

Initially, most of the studies were focused on the radiation effects on memories due to their higher occurrence and therefore higher impact on the functionality of the systems. Only after nearly 10 years, since the first observation of upsets in satellites by Binder et al. [3], the transient effects were observed in combinational logic circuits by May et al. [6]. Then, several works during the 1990s started to examine the anomalies in the combinational part of logic circuits and it was getting more attention from the radiation effects research community [7]. It was in the work developed in [8] which reported that radiation-induced transients could propagate and upset memory elements such as the latch gates. Though the transient effects were observed since 1984, the term *single-event transient* (SET) was only first adopted in 1990, by Newberry et al. [9]. Historically, SEUs have been vastly studied in the literature, while SETs were not given as much importance due to the intrinsic masking effects of combinational logic circuits [7]. However, the transistor scaling, reduced logic data path depth, and increased operating frequencies have attenuated the masking capability of logic circuits at advanced technology nodes [10–13]. Accordingly, several works started the development of radiation hardening techniques and mitigation schemes to reduce the impact of soft errors, i.e., both SEU and SET.

Although early research in radiation effects was predominantly focused on space and military systems, there has been a growing awareness of the impact on ground-level applications. For example, autonomous vehicles are increasingly susceptible to transient radiation effects, which could result in critical malfunctions if appropriate fault-tolerant measures are not implemented to enhance their reliability [45]. As discussed in the previous chapter, particle accelerators present an especially hostile environment for electronics [46]. A notable instance occurred within CERN's super proton synchrotron (SPS) in 2021, where multiple radiation-induced failures within the injection chain of the large hadron collider (LHC) significantly impacted the overall availability of the accelerator complex [47]. These failures were caused by data corruption in memory elements, i.e., SEU events, in a programmable logic controller (PLC)-based system.

Figure 2.1 illustrates the number of radiation to electronics (R2E) beam dumps that occurred during the large hadron collider's (LHC) physics runs in 2018 and 2024, plotted against the integrated luminosity at the compact muon solenoid (CMS) experiment. In 2018, a substantial number of failures were observed, particularly in power converters and quench protection systems, as a consequence of increased radiation levels caused by adjustments in the nominal operational settings of a collimator in the LHC ring [46].

To improve the overall availability of the accelerator complex, especially with the anticipated increase in integrated luminosity during the high-luminosity LHC (HL-LHC) era, a target failure rate of 0.1 R2E events per fb^{-1} has been set, as depicted in Fig. 2.1. This goal emphasizes the necessity of rigorous radiation monitoring and the comprehensive qualification of electronic systems deployed in the accelerator, ensuring adherence to a robust radiation hardness assurance (RHA) protocol to mitigate radiation-induced disruptions effectively.

Therefore, whether for space systems or ground-level applications, understanding the fundamental mechanisms of radiation effects is essential for designing

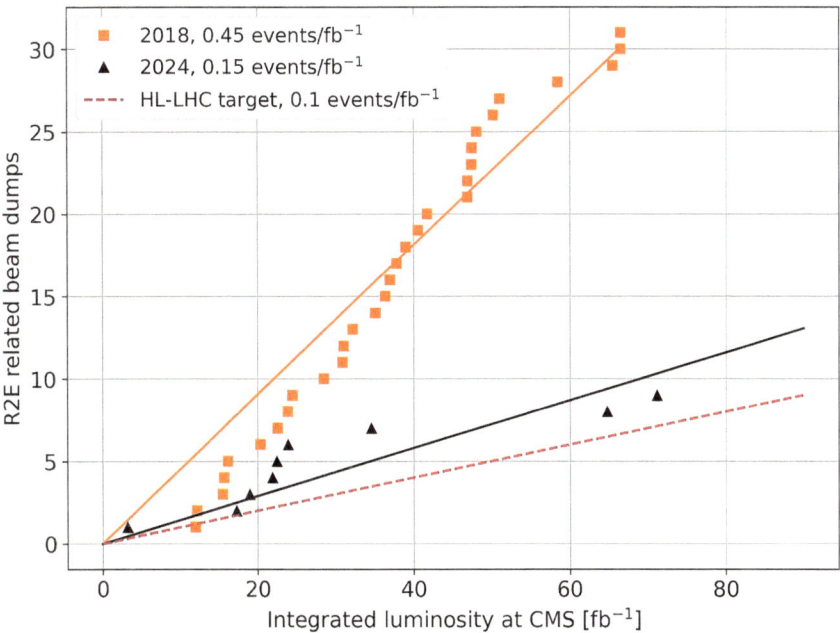

Fig. 2.1 Radiation-to-electronics (R2E)-induced beam dumps during LHC physics run in 2018 and 2014, as a function of the integrated luminosity at the compact muon solenoid (CMS) experiment. The R2E event rate for the high-luminosity large hadron collider (HL-LHC) period is also depicted in dashed line

reliable systems and enhancing their operational availability. In this chapter, the foundational concepts of single-event effects (SEE) in digital circuits will be introduced. These concepts will provide the necessary knowledge for addressing radiation-induced challenges in a wide range of applications, from space exploration to terrestrial technologies.

2.2 Single-Event Upset (SEU)

As mentioned previously, the SEUs are characterized by a single particle strike in memory elements leading to bit flips and consequently data corruption. To illustrate how a single particle can induce a SEU, Fig. 2.2 contains the gate representation of the basic element that composes a traditional Static Random Access Memory (SRAM) circuit topology, the cross-coupled inverters. This architecture is the core circuit of a traditional SRAM memory design in which additional access transistors (not shown in the figure) are used to read and write the logic states, i.e., the bit information.

This positive feedback circuit architecture is responsible for holding the bit information, therefore, working as a memory element. If a particle strikes any of

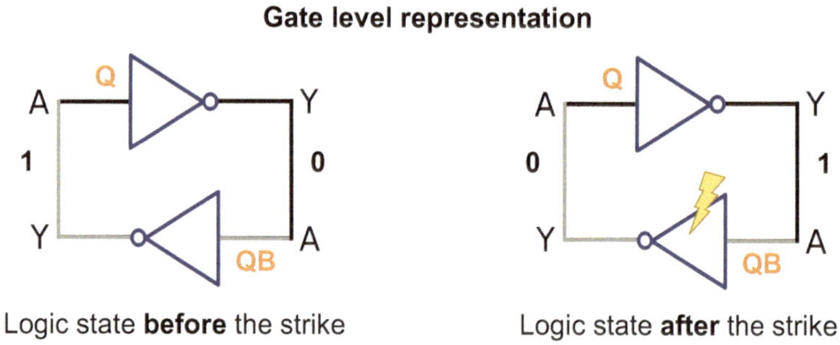

Gate level representation

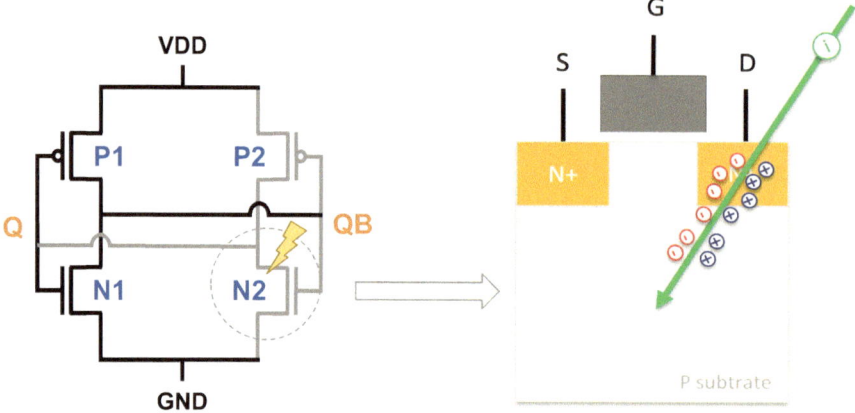

Logic state **before** the strike   Logic state **after** the strike

Fig. 2.2 Gate level representation of the Single-Event Upset (SEU) in a cross-coupled inverter pair

Fig. 2.3 Transistor-level representation of the single-event upset (SEU) in a cross-coupled inverter pair. Particle strikes at the drain region of the off-state N2 transistor and deposits energy via ionization

these inverter gates and changes its output signal, the feedback mechanism will hold the incorrect signal and then change the bit information storage in the circuit as shown in Fig. 2.2. To observe such effects, an energetic particle needs to hit next to one of the *off*-state transistors in the circuit (as shown in Fig. 2.3) and deposit sufficient energy. By doing so, the induced electron-hole pairs created by ionization can be collected by the reversed-biased p-n junction of the transistor, and a transient current is observed in its drain terminal.

If the amplitude and pulse width of this transient current are sufficiently large, in other words, if enough energy is deposited by the incident particle in the right place of the device, the output signal of the affected inverter will change and, consequently, the stored data will be corrupted, as illustrated in Fig. 2.2. Accordingly, the minimum collected charge necessary to change the output signal of the circuit is related to its nodal capacitance, and it is usually known as the **critical charge** (Q_{crit}) of the circuit. Thus, the sensitivity of such circuit depends on the node capacitance of its internal transistors which are responsible for the logic state retention. Accordingly, Q_{crit} can be expressed with the following simplified equation where C corresponds to the nodal capacitance and V_{DD} the supply voltage of the circuit:

$$Q_{crit} \sim 2V_{DD}C \qquad (2.1)$$

This is a simplified formula widely used to understand the implications of circuit design and to provide useful insights of the impact of transistor technology on the susceptibility to energetic particles. Together with the threshold Linear Energy Transfer (LET), as described in Chap. 1, the critical charge Q_{crit} is a very important parameter used to characterize components and estimate their SEE rate.

> **Critical charge** Q_{crit} of a circuit is widely used as a measurement of its sensitivity to SEE. In current state-of-the-art technologies, Q_{crit} is expected to be lower than $1fC$.

Besides the critical charge, another very important concept used to define the SEE sensitivity of a component is the *SEE cross-section* σ_{SEE}, which is a measure of the probability of a SEE to occur in a device. For a given LET, the event cross-section σ_{SEE} is calculated based on the number of observed events N_{SEE} under a given particle fluence ϕ as shown in Eq. 2.3:

$$\sigma_{SEE}(\text{LET}) = \frac{N_{SEE}}{\phi} \qquad (2.2)$$

The particle fluence, expressed in particles/cm^2, is the integrated number of particles passing through a unit area perpendicular to the beam direction. The sensitivity of a component is normally expressed in cross section as a function of the particle LET (or particle energy in the case of protons and neutrons). For ion-induced events, the energy deposition is dependent on the angle of incidence and, therefore, if the incident angle is not perpendicular to the device, the *effective LET* should be used instead:

$$\text{LET}_{\text{eff}} = \frac{LET_0}{\cos \theta} \qquad (2.3)$$

$$\sigma_{SEE}(\text{LET}_{\text{eff}}) = \frac{N_{SEE}}{\phi \cdot \cos \theta} \qquad (2.4)$$

Given the feedback mechanism of such architecture, two additional effects are observed at the circuit level, and they are essential for the failure analysis: the *transient propagation delay* and the *restoring current*. As shown in Fig. 2.4, the radiation-induced transient pulse observed in the output signal of the struck inverter

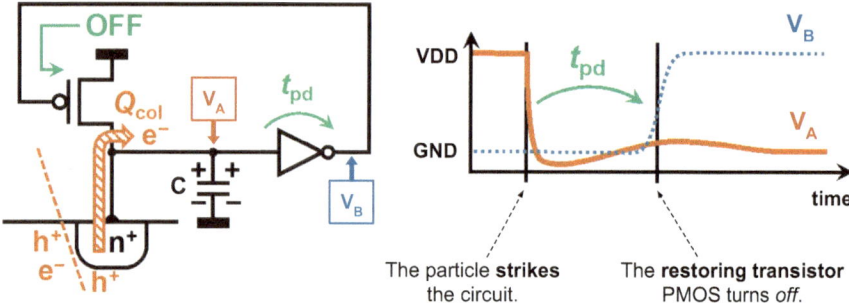

Fig. 2.4 Representation of the propagation delay of the transient pulse and the restoring transistor in the event of a particle strike (Adapted from [16])

propagates to the output of the second inverter with a propagation delay T_{pd} which is an intrinsic characteristic of the designed circuit considering a given technology node. While the transient pulse propagates to the input of the second inverter, the PMOS transistor in the first inverter is still *on*, and therefore it provides a restoring current that counteracts the radiation-induced transient current. The strength of the restoring transistor, i.e., its feature size, determines the resulting transient pulse observed in the circuit node. For instance, if the p-type transistor (PMOS device) in Fig. 2.4 is faster in restoring the output voltage than the propagation delay T_{pd}, no SEU will be observed in this memory cell. In addition to the reduced nodal capacitance, in the deeply scaled transistor technologies, the propagation delay tends to be smaller in every new generation [14, 15]. As a result, if no proper hardening approach is considered when adopting advanced technologies, a higher sensitivity to radiation effects should be expected.

Another issue observed when adopting deeply scaled technologies is the increase of the *charge sharing effect*, discussed in the previous chapter, that can lead to multiple-cell upset (MCU). In other words, within a single particle hit, multiple memory cells are upset due to their close proximity in the physical layout and the carrier diffusion mechanism through the substrate. Figure 2.5 illustrates the impact of scaling in a memory array and the consequent MCU phenomenon. When the affected cells correspond to the bits from the same logical word in the memory, they are called multiple-bit upset (MBU). The MBU occurrence has an important implication in the efficiency of error correcting codes (ECC) [17] where most of the techniques are designed to correct only single-bit upsets (SBUs). In order to preserve the data integrity considering MCUs, the ECC techniques would require a large number of redundant bits increasing the memory design complexity and the area overhead.

Another approach widely used along with the ECC techniques is the bit interleaving [18]. In a interleaving scheme, the memory cells are placed in such a way that the adjacent cells correspond to bits of different logical words. Therefore, if the memory array in Fig. 2.5 adopts bit interleaving, the MCU event would not lead to a MBU and the ECC techniques would be able to correct the SBU from

1-bit memory cell

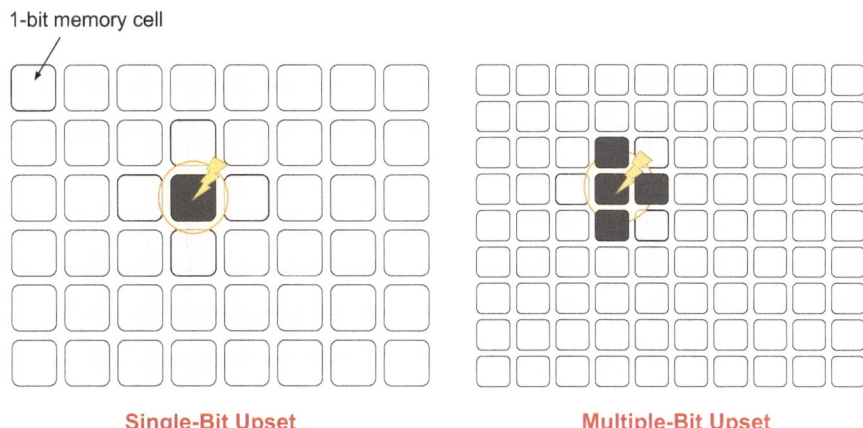

Single-Bit Upset   **Multiple-Bit Upset**

Fig. 2.5 Single-Bit Upset (SBU) vs. Multiple-Bit Upset (MBU). Due to transistor scaling and the consequent higher cell density, for the same particle hit, a larger number of memory cells is affected in deeply scaled technologies

the different affected words. Although the increased cell density in memories has shown an increased sensitivity to MCUs and MBUs in the latest planar devices, the memories using three-dimensional devices such as FinFET have shown an improved resilience due to the higher substrate doping profiles [19, 20]. In order to suppress the short-channel effects (SCEs) such as increased off-state leakage current, higher substrate doping levels are used in FinFET devices which limit the carrier mobility and therefore reduces the charge sharing effect and the multiple node charge collection. However, it was also shown that in both planar and FinFET SRAMs, the MBU susceptibility has a strong angle dependence where the device orientation could be critical in determining the efficiency of ECC techniques [21].

> **Charge sharing** is a predominantly negative effect that increases the sensitivity of advanced technologies. However, design techniques can take into consideration to propose hardening strategies as shown in the next chapters.

2.3   Single-Event Functional Interruption (SEFI)

The consequences of an SEU depend on the nature of the corrupted information and the timing of the event. In simpler devices, the error might go unnoticed. However, for complex systems like processors and Field-Programmable Gate Arrays (FPGAs) devices, an SEU can lead to critical misbehavior. When an SEU occurs in a control or state-holding section of a complex device, it can trigger a more severe consequence known as a single-event functional interruption (SEFI). A SEFI manifests as a complete cessation of operation in the affected circuit, often resulting in a system reset or lockup. Imagine a tiny glitch causing your computer to freeze entirely—that is akin to a SEFI in electronics. Unlike some soft errors that might cause transient hiccups, SEFIs are typically detectable due to the complete halt in operation.

SEFIs are most prevalent in devices with integrated control or state sections, such as modern memories, processors, FPGAs, and Application-Specific Integrated Circuits (ASICs). This is because these devices rely heavily on accurate logic states and control signals, and any disruption caused by an SEU can have a significant impact on functionality. Understanding SEFIs requires familiarity with three key concepts in reliability in electronics:

- **Fault:** A physical defect within a device that can trigger errors under certain conditions. In the context of radiation effects, faults can arise from disruptions in critical circuitry due to ionization or displacement damage caused by particle interactions. For instance, a radiation-induced transient in an off-state transistor is a fault.
- **Error:** An error is the manifestation of a fault and implies a deviation from the intended behavior of a system, for example, a bit flip in a memory element that corrupts critical data.
- **Failure:** A failure occurs when a system or component cannot fulfill its intended function. In the context of SEFI, the failure is marked by a catastrophic loss of functionality in the affected circuit, potentially compromising the integrity of the entire system.

SEUs are known as soft errors because they do not present permanent damage, and they are often silent, i.e., they alter the state of a memory cell without system detection. These undetected errors can propagate through the system, potentially leading to failures. However, if a soft error occurs in a control unit or critical part of the system, it can disrupt operation in a detectable way, allowing the system to take corrective measures and potentially avoid a complete failure. SEFIs, in essence, are a type of fault that manifests as a detectable error, often with the potential for self-correction (through a system reset). This differentiates them from potentially silent soft errors.

2.4 Single-Event Transient (SET)

The basic mechanisms observed in SEUs and introduced in the last subsection also apply to SETs in logic circuits. In fact, one of the main differences between SETs and SEUs is the type of circuit affected by particle interaction. Overall, digital logic circuits can be classified into two groups: *combinational logic* and *sequential logic* circuits. The combinational logic circuit implements a Boolean logic function which only depends on the actual set of inputs of the circuit. For instance, the NOT gates

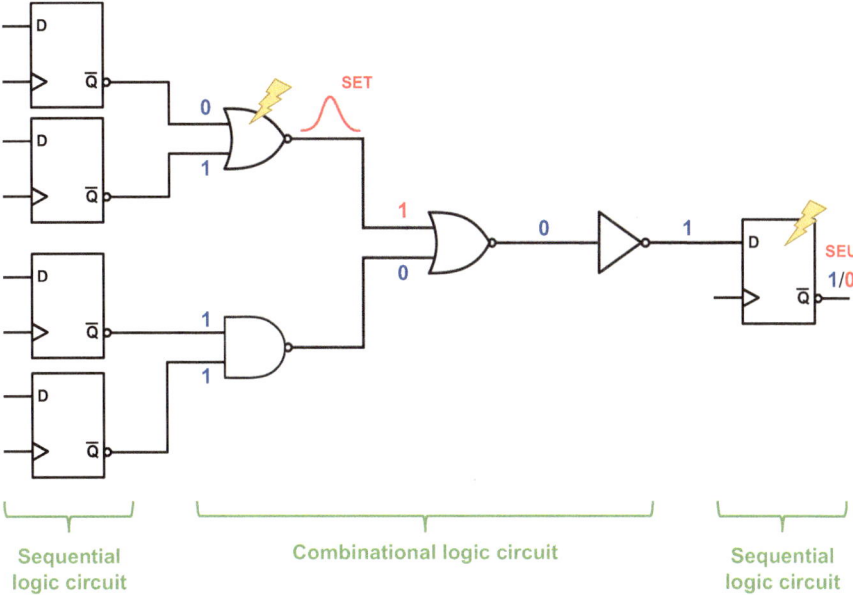

Fig. 2.6 Single-event transient (SET) definition vs. single-event upset (SEU)

(inverters) which are used to implement the logical negation, and the AND/OR gates to implement the logical conjunction/disjunction are very commonly used combinational logic gates. On the other hand, in a sequential logic architecture, the logic implementation depends not only on the actual input of the circuit but also on the previous inputs and outputs. Therefore, *"sequential"* refers to the sequence of information which introduces the notion of storage. A simple way to implement this sequential mechanism is by using positive feedback as shown previously in Fig. 2.2. Besides the SRAM design, this feedback approach is widely used to design different sequential logic circuits used as storage elements such as registers, latches, and flip-flops. In Fig. 2.6, an example circuit illustrates the sequential logic gates (in this case, the flip-flops) and the combinational logic gates such as the NAND (not AND), the NOR (not OR), and inverter.

In contrast to SEU which occurs in the sequential logic part of the circuit, the SET occurs in the combinational one. Therefore, the transient pulse is generated in the output of the logic gate, and it can propagate through the data path until it is latched by a memory element and corrupts the stored bit. However, for this SET pulse to upset a memory element at the end of its data path, it needs to surpass three basic masking capabilities inherent in any combinational circuit: logical masking, electrical masking, and latching-window masking (also known as temporal masking).

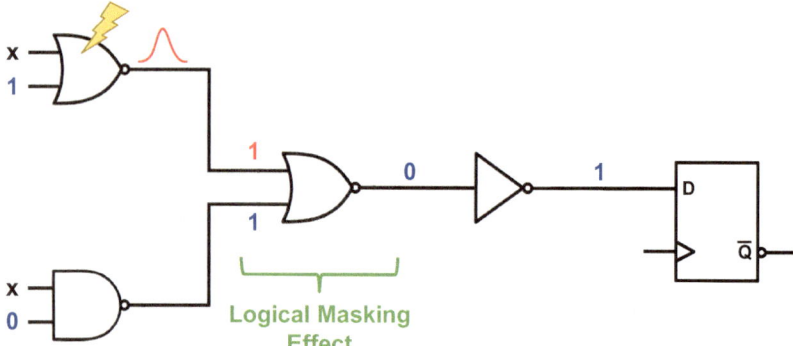

Fig. 2.7 Illustration of the logical masking effect of a SET event in a two-input NOR gate within a block of combinational circuit

2.4.1 Logical Masking Effect

Combinational circuits provide the logical masking effect when the SET event occurs in a logic gate where its logic output does not determine the output signal of the subsequent logic stage. For instance, a two-input NOR gate has its output determined whenever one of its input signals is evaluated to 1, i.e., whenever one input signal is at logic 1, the output evaluates to logic 0. This phenomenon can be better understood by analyzing the block of combinational logic in Fig. 2.7. A SET event occurs in the first NOR gate, in which the output signal initially was evaluated to logic 0. The SET pulse propagates to the next logic stage, which is also a NOR gate. However, this logic stage has already been evaluated to logic 0 due to the input signal provided by the NAND gate. Since the output of the second NOR gate has already been determined by one of its inputs, the SET pulse is not able to change it; hence, it is logically masked and cannot propagate to the next logic stage and reach a memory element, for instance. Although this mechanism is effective, recent technologies have shown a reduction in the logic depth of combinational circuits, thus logical masking effect has been reduced [22]. Nevertheless, circuit designers can promote the logical masking effect by introducing more basic logic gates in the design instead of using complex logic gates as shown in [23].

2.4.2 Electrical Masking Effect

The electrical masking effect is another phenomenon that can occurs in a combinational circuit and prevents the propagation of a SET pulse. Due to electrical losses, a SET pulse suffers from magnitude and amplitude attenuation, and it might not be able to propagate to a memory element as observed in Fig. 2.8. The initial SET pulse has its waveform affected by each stage of logic, vanishing near the memory element.

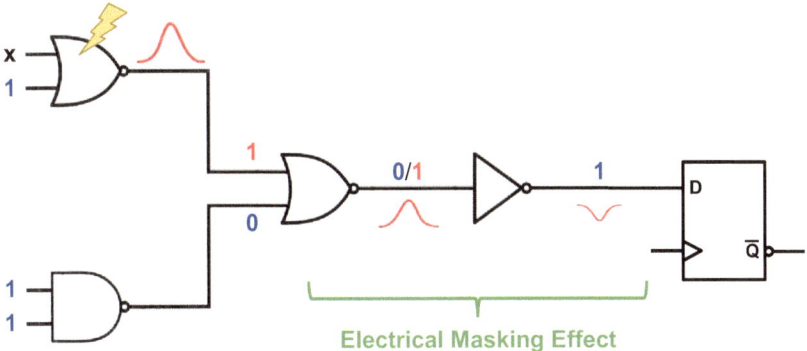

Fig. 2.8 Illustration of the electrical masking effect of a SET event due to electrical losses in a logic path

In this case, the propagated SET pulse did not have sufficient amplitude to upset the memory element due to the electrical masking effect. However, it was shown that not only the transient pulse can suffer from attenuation, but it can also experience a broadening effect, the so-called *propagation-induced pulse broadening* (PIPB) [24, 25]. Similarly to the SEU, the pulse width of the SET depends on the restoring current of the struck circuit and its capacitive load (fan-out). Larger capacitance can lead to increase in the critical charge; however, it can lead to pulse broadening due to the longer time to restore the output voltage [26]. Due to its complexity, the PIPB effect is difficult to be evaluated and, therefore, it is a significant issue when considering hardening methodologies for SET mitigation [12, 27].

2.4.3   Latching-Window Masking Effect

In the end, if the SET pulse has not been masked logically or electrically, it might still be masked by the latching window of a memory element. This window is composed by the *setup time* (T_{setup}) and the *hold time* (T_{hold}) around the edge of the clock signal of a flip-flop circuit. If the SET pulse does not arrive during this latching window, it will not be able to induce a bit upset, i.e., a change in the stored bit value. Figure 2.9 illustrates this phenomenon. Due to the high operating clock frequencies in advanced technologies, the latching-window effect is expected to be reduced given the short T_{setup} and T_{hold} of FF designs [28].

In summary, the particle strike must induce a SET pulse with sufficient amplitude and duration to propagate through an open logic path and reach a memory element during a clock pulse, enabling the latching of the input value. Thus, as the clock frequency increases and the supply voltage reduces, the probability of a SET pulse to be latched by a memory element increases [29].

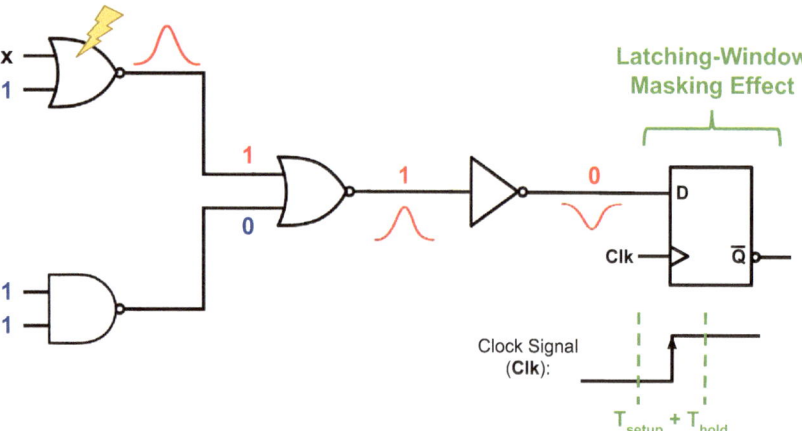

Fig. 2.9 Illustration of the latching-window masking effect of a SET pulse by a flip-flop (FF) circuit

Masking effects are naturally present in digital logical circuits and their effectiveness depends on several factors such as the transistor technology, circuit design, and operation parameters (clock frequency, supply voltage, temperature, and so on).

2.5 Single-Event Latchup (SEL)

Since the 1970s, the latchup mechanism has been a very well-known reliability issue observed in bulk Complementary Metal-Oxide Semiconductor (CMOS) technology due to the parasitic *pnpn* structure inherently present in this technology as shown in Fig. 2.10 [30–32]. In nominal operation, this structure composed of parasitic bipolar junction transistors (BJTs) is under high impedance, and therefore, it does not interfere in the circuit operation. However, these transistors can be activated externally by (1) electrical stress, known as *electrical latchup* or *transient-induced latchup* or (2) induced by particle radiation, known as *single-event latchup* (SEL). Different from the soft errors, such as the SEU and SET discussed previously, SEL can have a destructive effect depending on the duration and magnitude of the induced parasitic current.

This phenomenon occurs when the particle interaction within a circuit leads to the activation of these parasitic BJTs and therefore the creation of a low-impedance path between both power rails, the power supply, and ground. Once this connection is established, a high current is observed in the circuit which can permanently

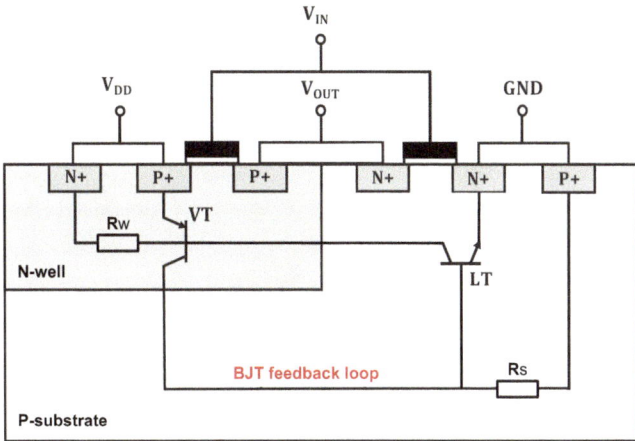

Fig. 2.10 Cross-sectional view of a typical CMOS-based inverter design and its inherent parasitic bipolar transistors (Adapted from [33])

damage the component. To restore the correct functionality of the circuit and prevent its permanent damage, a power cycle is necessary. In Fig. 2.10, the cross-sectional view of a typical CMOS-based inverter design is shown in which the parasitic BJTs are also illustrated. In the literature, these BJTs are also often known as vertical transistor (VT) and lateral transistor (LT). The resistors R_W and R_S correspond to the well and substrate resistance, respectively. The placement of the well-taps and the doping profile of the CMOS structure determines the value of these resistances and therefore the electrical characteristics required to observe a sustainable SEL.

Notice that these BJTs are not designed to be active during the nominal operation of the circuit; however, they are formed due to the interplay between the different junctions, potentials, and nested wells within the CMOS structure itself. Due to their positive feedback connection, the BJTs will only turn *off*, and, the SEL will be extinguished when the supply voltage of the circuit is reduced to a level below the so-called *holding voltage* (V_{hold}). Figure 2.11 illustrates the case when a particle strike deposits enough charge to activate the feedback loop of BJTs leading to a sustained SEL (until the supply voltage is reduced under the V_{hold}) and the case in which the radiation-induced current is not sufficient to turn *on* the parasitic BJT loop. Besides the V_{hold}, another latchup criterion is related to the BJT gain of both lateral and vertical transistors. In order to establish the positive feedback, the product of their gains must exceed unity, i.e., $\beta_V \beta_L > 1$ [34].

Together with soft errors, the threat of SEL in electronic components has been a well-known and highly important concern for systems operating in the space environments [36–39]. Besides the circuit design characteristics, the SEL sensitivity of a given component also depends on the environmental temperature as proven in the literature [33, 35, 40–43]. The temperature variation modifies the intrinsic characteristics of the devices such as the carrier mobility and the threshold voltage. For example, carrier mobility decreases with the increase in temperature; therefore, the substrate/well resistance increases accordingly. As a consequence of such

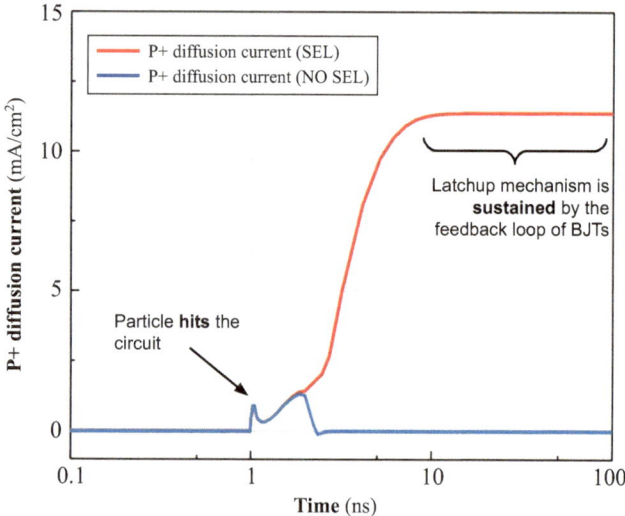

Fig. 2.11 Radiation-induced current in the p+ diffusion when the latchup mechanism is triggered (red) and when it is not (blue) (Adapted from [35])

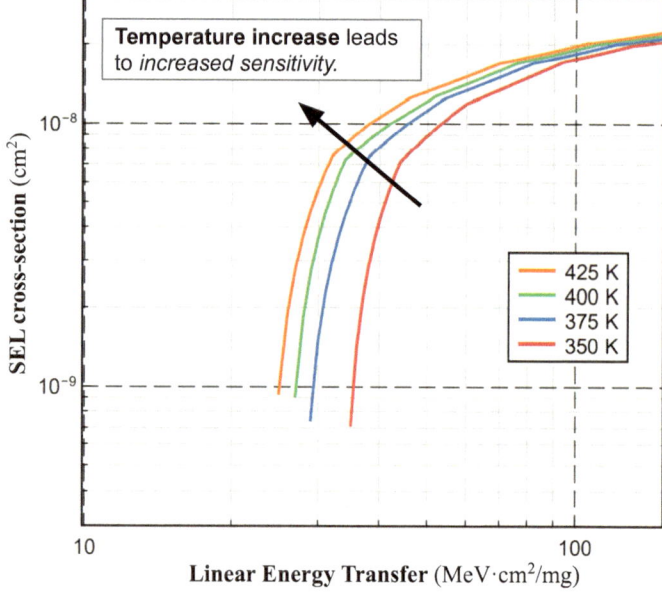

Fig. 2.12 Temperature effect on the SET cross-section curves of a 65 nm inverter design (Adapted from [35])

variation in the substrate and well resistances, the SEL triggering mechanism is also affected. In Fig. 2.12, the SEL sensitivity of an inverter gate designed in 65 nm bulk CMOS is shown as a function of particle LET and temperature of operation. The sensitivity is measured in terms of the SEL cross section which increases with temperature for both low and high LET values. Thus, an increase in temperature can not only increase the saturation SEL cross section but can also lower the threshold LET necessary to trigger the latchup. However, these relationships are not as simple as they might seem. For instance, the temperature dependence of carrier mobility is also a function of the doping concentration: the lower the concentration, the stronger the temperature dependence [44]. It is for this reason that the SEL susceptibility of a given circuit is highly dependent on the layout design itself, but also on the process technology. Therefore, in Chap. 4, the implications of process technologies and mostly design approaches on the overall SEE sensitivity of a circuit are discussed.

2.6 Summary

Radiation effects are no longer an exclusive concern for system designers targeting space or military applications. Due to the advancement of technology, low-energy particles present even at sea level are able to induce failure mechanisms in the device and circuit levels. In this chapter, the stochastic effects known as single-event effects (SEEs) were presented. The most well-known and highly important effects are the soft errors, i.e., the nondestructive effects named single-event upset (SEU) and single-event transient (SET). Their failure signature is strongly related to the data corruption either by a direct impact on the memory element themselves or by a transient pulse which is propagated through the data path and latched during the read mode.

Initially, the SETs in digital electronics showed less of a concern due to the inherent masking effect capability of combinational circuits. However, with the technology integration, the effectiveness of these masking effects has diminished, and a higher impact of SET is observed in today's electronic technology. Another very relevant failure mechanism in modern technology is the single-event latchup (SEL), a potentially destructive effect that if no action is taken, the electronic components can suffer permanent damage due to the very high current flow between the supply rails.

To characterize and investigate these effects in different systems, simulation codes are widely employed in conjunction with irradiation testing campaigns. Therefore, in the next chapter, we will discuss the physical models used to describe these phenomena and provide an overview of various simulation tools available in the literature. This understanding of SEEs and the associated analysis tools is essential for effectively assessing and mitigating their impact on electronic systems in diverse applications.

Highlights

- Single-event effects (SEEs) are stochastic effects caused by single particle interactions, and they can be destructive or nondestructive (soft errors).
- A single-event functional interruption (SEFI) is a manifestation of soft errors in complex devices, leading the component to reset, lockup or experience other types of malfunction.
- Among several factors, the SEE sensitivity of a circuit is dependent on the transistor technology, supply voltage, and nodal capacitance of transistors.
- The threshold LET and the critical charge of a component are important parameters to characterize and estimate its SEE rate.
- The multiple-node charge collection due to the charge sharing can pose a threat to the efficiency of error-correcting codes (ECCs) in advanced technologies.
- The masking effect capability of digital combinational circuit has been reduced in deeply scaled transistor technologies.
- The single-event latchup (SEL) is a potentially destructive SEE that can damage the component if the supply voltage is not reduced under the so-called holding voltage V_{hold}.

References

1. Robert Baumann. The impact of technology scaling on soft error srate performance and limits to the efficacy of error correction. In *Digest. International Electron Devices Meeting,*, pages 329–332. IEEE, 2002.
2. JT Wallmark and SM Marcus. Minimum size and maximum packing density of nonredundant semiconductor devices. *Proceedings of the IRE*, 50 (3): 286–298, 1962.
3. D Binder, EC Smith, and AB Holman. Satellite anomalies from galactic cosmic rays. *IEEE Transactions on Nuclear Science*, 22 (6): 2675–2680, 1975.
4. Timothy C May and Murray H Woods. A new physical mechanism for soft errors in dynamic memories. In *Reliability Physics Symposium, 1978. 16th Annual*, pages 33–40. IEEE, 1978.
5. CS Guenzer, EA Wolicki, and RG Allas. Single event upset of dynamic rams by neutrons and protons. *IEEE Transactions on Nuclear Science*, 26 (6): 5048–5052, 1979.
6. TC May, GL Scott, ES Meieran, P Winer, and VR Rao. Dynamic fault imaging of VLSI random logic devices. In *22nd International Reliability Physics Symposium*, pages 95–108. IEEE, 1984.
7. Veronique Ferlet-Cavrois, Lloyd W Massengill, and Pascale Gouker. Single event transients in digital CMOS—a review. *IEEE Transactions on Nuclear Science*, 60 (3): 1767–1790, 2013.
8. KJ Hass, RK Treece, and AE Giddings. A radiation-hardened 16/32-bit microprocessor. *IEEE Transactions on Nuclear Science*, 36 (6): 2252–2257, 1989.
9. DM Newberry, DH Kaye, and GA Soli. Single event induced transients in I/O devices: A characterization. *IEEE Transactions on Nuclear Science*, 37 (6): 1974–1980, 1990.
10. Stephen P Buchner. Single-event transients in fast electronic circuits. *NSREC Short Course, 2001*, 2001.

11. JM Benedetto, PH Eaton, DG Mavis, M Gadlage, and T Turflinger. Digital single event transient trends with technology node scaling. *IEEE Transactions on Nuclear Science*, 53 (6): 3462–3465, 2006.

12. Matthew J Gadlage, Jonathan R Ahlbin, Balaji Narasimham, Bharat L Bhuva, Lloyd W Massengill, Robert A Reed, Ronald D Schrimpf, and Gyorgy Vizkelethy. Scaling trends in set pulse widths in sub-100 nm bulk CMOS processes. *IEEE Transactions on Nuclear Science*, 57 (6): 3336–3341, 2010.

13. NN Mahatme, I Chatterjee, BL Bhuva, J Ahlbin, LW Massengill, and R Shuler. Analysis of soft error rates in combinational and sequential logic and implications of hardening for advanced technologies. In *2010 IEEE International Reliability Physics Symposium*, pages 1031–1035. IEEE, 2010.

14. Ali Khakifirooz and Dimitri A. Antoniadis. Mosfet performance scaling—part i: Historical trends. *IEEE Transactions on Electron Devices*, 55 (6): 1391–1400, 2008. doi: https://doi.org/10.1109/TED.2008.921017.

15. Kelin J. Kuhn. Considerations for ultimate CMOS scaling. *IEEE Transactions on Electron Devices*, 59 (7): 1813–1828, 2012. doi: https://doi.org/10.1109/TED.2012.2193129.

16. Daisuke Kobayashi. Basics of single event effect mechanisms and predictions.

17. Richard W Hamming. Error detecting and error correcting codes. *The Bell system technical journal*, 29 (2): 147–160, 1950.

18. Pedro Reviriego, Juan Antonio Maestro, and Catalina Cervantes. Reliability analysis of memories suffering multiple bit upsets. *IEEE Transactions on Device and Materials Reliability*, 7 (4): 592–601, 2007.

19. Yi-Pin Fang and Anthony S Oates. Characterization of single bit and multiple cell soft error events in planar and FinFET SRAMs. *IEEE Transactions on Device and Materials Reliability*, 16 (2): 132–137, 2016.

20. Balaji Narasimham, Vikas Chaudhary, Mike Smith, Liming Tsau, Dennis Ball, and B Bhuva. Scaling trends in the soft error rate of SRAMs from planar to 5-nm FinFET. In *2021 IEEE International Reliability Physics Symposium (IRPS)*, pages 1–5. IEEE, 2021.

21. Takashi Kato, Masanori Hashimoto, and Hideya Matsuyama. Angular sensitivity of neutron-induced single-event upsets in 12-nm FinFET SRAMs with comparison to 20-nm planar SRAMs. *IEEE Transactions on Nuclear Science*, 67 (7): 1485–1493, 2020.

22. Selahattin Sayil. *Soft error mechanisms, modeling and mitigation*. Springer, 2016.

23. Y. Q. Aguiar, Frédéric Wrobel, J-L Autran, Paul Leroux, Frédéric Saigné, AD Touboul, and Vincent Pouget. Impact of complex-logic cell layout on the single-event transient sensitivity. *IEEE Transactions on Nuclear Science*, 66 (7): 1465–1472, 2019.

24. Paul E Dodd, Marty R Shaneyfelt, James A Felix, and James R Schwank. Production and propagation of single-event transients in high-speed digital logic ICS. *IEEE Transactions on Nuclear Science*, 51 (6): 3278–3284, 2004.

25. V Ferlet-Cavrois, P Paillet, D McMorrow, N Fel, J Baggio, S Girard, O Duhamel, JS Melinger, M Gaillardin, JR Schwank, et al. New insights into single event transient propagation in chains of inverters—evidence for propagation-induced pulse broadening. *IEEE Transactions on Nuclear Science*, 54 (6): 2338–2346, 2007.

26. Gilson Wirth, Fernanda L Kastensmidt, and Ivandro Ribeiro. Single event transients in logic circuits—load and propagation induced pulse broadening. *IEEE Transactions on Nuclear Science*, 55 (6): 2928–2935, 2008.

27. V Ferlet-Cavrois, Vincent Pouget, D McMorrow, JR Schwank, N Fel, Fabien Essely, RS Flores, P Paillet, M Gaillardin, Daisuke Kobayashi, et al. Investigation of the propagation induced pulse broadening (PIPB) effect on single event transients in SOI and bulk inverter chains. *IEEE Transactions on Nuclear Science*, 55 (6): 2842–2853, 2008.

28. Premkishore Shivakumar, Michael Kistler, Stephen W Keckler, Doug Burger, and Lorenzo Alvisi. Modeling the effect of technology trends on the soft error rate of combinational logic. In *Proceedings International Conference on Dependable Systems and Networks*, pages 389–398. IEEE, 2002.

29. D Munteanu and J-L Autran. Modeling and simulation of single-event effects in digital devices and ICS. *IEEE Transactions on Nuclear science*, 55 (4): 1854–1878, 2008.

30. JF Leavy and RA Poll. Radiation-induced integrated circuit latchup. *IEEE Transactions on Nuclear Science*, 16 (6): 96–103, 1969.
31. BL Gregory and BD Shafer. Latch-up in CMOS integrated circuits. *IEEE Transactions on Nuclear Science*, 20 (6): 293–299, 1973.
32. DB Estreich, A Ochoa, and RW Dutton. An analysis of latch-up prevention in CMOS IC's using an epitaxial-buried layer process. In *1978 International Electron Devices Meeting*, pages 230–234. IEEE, 1978.
33. S Guagliardo, Frédéric Wrobel, YQ Aguiar, J-L Autran, Paul Leroux, Frédéric Saigné, Vincent Pouget, and AD Touboul. Single-event latchup sensitivity: Temperature effects and the role of the collected charge. *Microelectronics Reliability*, 119: 114087, 2021.
34. Ronald R Troutman. Latchup in CMOS technologies. *IEEE Circuits and Devices Magazine*, 3 (3): 15–21, 1987.
35. S. Guagliardo, Frederic Wrobel, Ygor Q. Aguiar, JL Autran, P Leroux, Frédéric Saigné, Vincent Pouget, and Antoine D Touboul. Effect of temperature on single event latchup sensitivity. In *2020 15th Design & Technology of Integrated Systems in Nanoscale Era (DTIS)*, pages 1–5. IEEE, 2020.
36. W. A. Kolasinski, J. B. Blake, J. K. Anthony, W. E. Price, and E. C. Smith. Simulation of cosmic-ray induced soft errors and latchup in integrated-circuit computer memories. *IEEE Transactions on Nuclear Science*, 26 (6): 5087–5091, 1979. doi: https://doi.org/10.1109/TNS.1979.4330278.
37. K. Soliman and D. K. Nichols. Latchup in cmos devices from heavy ions. *IEEE Transactions on Nuclear Science*, 30 (6): 4514–4519, 1983. doi: https://doi.org/10.1109/TNS.1983.4333163.
38. Yves Moreau, Helene de la Rochette, Guy Bruguier, Jean Gasiot, Frederic Pelanchon, Christophe Sudre, and Robert Ecoffet. The latchup risk of CMOS-technology in space. *IEEE transactions on nuclear science*, 40 (6): 1831–1837, 1993.
39. AH Johnston. The influence of VLSI technology evolution on radiation-induced latchup in space systems. *IEEE Transactions on Nuclear Science*, 43 (2): 505–521, 1996.
40. WA Kolasinski, R Koga, E Schnauss, and J Duffey. The effect of elevated temperature on latchup and bit errors in CMOS devices. *IEEE Transactions on Nuclear Science*, 33 (6): 1605–1609, 1986.
41. AH Johnston, BW Hughlock, MP Baze, and RE Plaag. The effect of temperature on single-particle latchup. *IEEE Transactions on nuclear science*, 38 (6): 1435–1441, 1991.
42. Cheryl J Marshall, Paul W Marshall, Raymond L Ladbury, Augustyn Waczynski, Rajan Arora, Roger D Foltz, John D Cressler, Duncan M Kahle, Dakai Chen, Gregory S Delo, et al. Mechanisms and temperature dependence of single event latchup observed in a CMOS readout integrated circuit from 16–300 k. *IEEE Transactions on Nuclear Science*, 57 (6): 3078–3086, 2010.
43. A Al Youssef, L Artola, S Ducret, G Hubert, and F Perrier. Investigation of electrical latchup and SEL mechanisms at low temperature for applications down to 50 k. *IEEE Transactions on Nuclear Science*, 64 (8): 2089–2097, 2017.
44. SM Sze and Kwok K Ng. Physics of semiconductor devices, John Wiley & Sons. *New York*, 68, 1981.
45. Lurada, Leonardo, Cavagnero, Niccolò, Santos, Fernando Fernandes Dos, Averta, Giuseppe, Rech, Paolo, and Tommasi, Tatiana. Transient Fault Tolerant Semantic Segmentation for Autonomous Driving, arXiv preprint arXiv:2408.16952, 2024.
46. Aguiar, Ygor, Apollonio, Andrea, Lerner, Giuseppe, Cerutti, Francesco, Sabaté-Gilarte, Marta, Danzeca, Salvatore, Prelipcean, Daniel, and García Alía, Ruben. Radiation to electronics impact on CERN LHC operation: Run 2 overview and HL-LHC outlook, 2021.
47. Aguiar, Ygor, Apollonio, Andrea, Lerner, Giuseppe, Cecchetto, Matteo, Potoine, Jean-Baptiste, Brucoli, Matteo, Danzeca, Salvatore, García Alía, Ruben, Biłko, Kacper, Zimmaro, Alessandro, and others. Implications and mitigation of radiation effects on the CERN SPS operation during 2021, *JACoW IPAC*, 2022, 740–743, 2022.

Chapter 3
Single-Event Effect Prediction Methodologies

3.1 Modeling and Prediction Tools

In addition to experimental activities, the utilization of modeling and simulation has long played a vital role in investigating physical phenomena, particularly in the field of electronics, where it has been instrumental in studying the behavior of Metal-Oxide Semiconductor (MOS) transistors [1, 2]. With the increasing complexity of very-large-scale integration (VLSI) systems in each new technology generation, simulation studies have become indispensable for verifying and aiding in the development of such circuits. In this context, Monte Carlo simulation tools have emerged as a robust approach for exploring radiation effects on electronics [3]. Numerous studies in the literature have leveraged simulations to investigate radiation effects on electronics, offering an alternative to time-consuming and expensive radiation campaigns [4–11].

In Table 3.1, a non-exhaustive list of simulation tools dedicated to model and study radiation effects on electronics is presented. Further details adopted in each tool can be found in their respective reference. In the simulations developed in this book, the MC-Oracle tool [6] is used to account for the energy deposition and charge collection in our SEE predictive methodology.

While simulation tools cannot completely replace the need for experimental data, they offer valuable insights into radiation effects by providing useful pre- and post-irradiation information. By employing modeling and simulation, researchers can gain a better understanding of the various mechanisms that occur at the circuit and component levels, thereby maximizing the outcomes of an irradiation campaign. Additionally, simulation enables the testing of hypothetical devices or conditions that may be challenging to reproduce experimentally or during an irradiation campaign.

Mixed-mode technology computer-aided design (TCAD) simulations have been vastly used to understand the main mechanisms in SEEs on electronics. However, Monte Carlo (MC) simulation codes can have a computation time several orders

© The Author(s) 2025
Y. Quadros de Aguiar et al., *Single-Event Effects, from Space to Accelerator Environments*, https://doi.org/10.1007/978-3-031-71723-9_3

Table 3.1 List of simulation tools dedicated to study radiation effects on electronics

Ref.	Tool	Institution	Radiation effects
[12]	DASIE	AIRBUS	SEU
[13]	SEMM-2	IBM	SEU
[14]	MRED	Vanderbilt University and NASA/GSFC	SEU
[6]	MC-ORACLE	Université de Montpellier	SEU/SET
[15]	TIARA-G4	Aix-Marseille University and STMicroelectronics	SEU/SET
[16]	FLUKA	CERN and INFN	SEU
[5]	MUSCA-SEP3	ONERA	SEU/SET/SEL
[17]	G4SEE	CERN	SEE

of magnitude lower than TCAD simulations [3, 10]. In addition, the randomness and stochastic nature of particle interaction with matter is a perfect fit for Monte Carlo (MC) simulations. Accordingly, a diverse number of models based on the MC method have been proposed to estimate and predict the radiation robustness of electronics [4–7, 10]. Unlike deterministic models, the MC method utilizes random sampling and statistical modeling to approximate solutions for stochastic problems, such as in particle physics.

3.2 SEE Triggering Criterion

The key ingredient of SEE simulation tools is the triggering criterion, which states whether a given particle in given conditions is able to trigger an SEE. Here we present two widely used criterions: the rectangular parallelepiped (RPP) and the drift-diffusion-collection.

3.2.1 Rectangular Parallelepiped (RPP) Criterion

Historically, the well-known *rectangular parallelepiped (RPP) analytical model* [18], also known as *chord-length model*, has been vastly used to analyze and predict the radiation response of electronic components [19]. In this approach, the device is assumed to have a well-defined sensitive volume (SV) in the form of a rectangular parallelepiped as shown in Fig. 3.1. The ionization track path of interest for the radiation effect in the circuit is determined by the depth of the SV of the device and the particle's incidence angle, θ. In this model, it is assumed that charge collection induced by diffusion from particles striking outside the RPP is negligible and thus not considered.

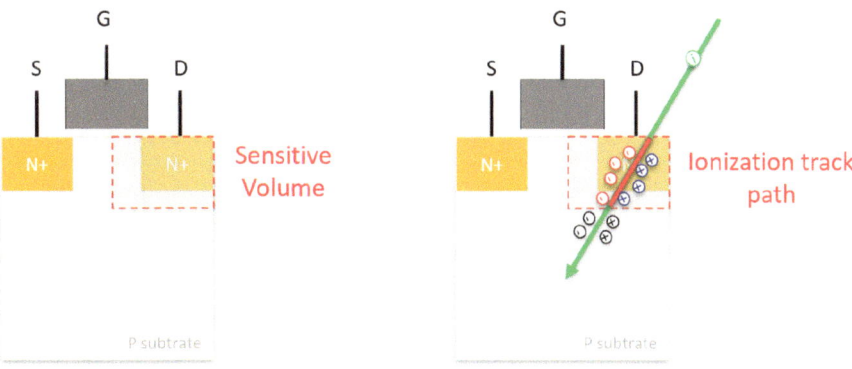

Fig. 3.1 Definition of a sensitive volume and ionization track path in the rectangular paral-lelepiped (RPP) analytical model

> The **RPP criterion** states that an SEE occurs when sufficient charge (critical charge) is generated inside a volume surrounding the drain of the OFF transistor (sensitive volume). Both the critical charge and the sensitive volume are characteristic of the considered technology.

Considering the small dimensions of transistors and their sensitive volume, the linear energy transfer (LET) is assumed to be constant over the ionization path. Therefore, the deposited charge within the SV can be calculated by the product of the LET value and the ionization path length, l. As shown in the Chap. 1, the deposited charge Q_D can be calculated using Eq. 1.6, reproduced here for reference:

$$Q_D = \frac{q \cdot \text{LET} \cdot \rho \cdot l}{E_{ehp}} \tag{3.1}$$

Considering the target material in electronic components is silicon, the silicon density ρ_{Si} is 2.32 g $\cdot$ cm^{-3}, the ionization energy E_{ehp} in Si is approximately 3.6 eV/ehp, and the elementary charge q is $1.6 \cdot 10^{-19} C$. In the case of a normal incidence, the ionization path length l is equal to the sensitive volume depth, d. Thus, from Eq. 3.1, the collected charge Q_C in the RPP model can be estimated as described in Eq. 3.2:

$$Q_C \text{ [pC]} = 10.35 \cdot \text{LET [MeV.cm}^2\text{/mg]} \cdot d \text{ [nm]} \tag{3.2}$$

Accordingly, if sufficient charge is deposited inside the SV, i.e., if the collected charge Q_C is superior to the critical charge Q_{crit}, an SEE is assumed to be observed in the circuit. Despite its popularity and its widespread use, the RPP model has turned out to be inadequate when used in advanced technology due to the complex geometry of transistors, the small sensitive volumes, and the close proximity of devices. Emerging effects such as charge sharing due to multiple node collection and parasitic bipolar amplification (PBA) have limited the application of the RPP approach in some cases.

> The **RPP analytical model** is a first-order approximation used to predict SEE rates based on some assumptions on the sensitive volume of a given component. Its application is limited when considering deeply scaled device technologies.

One possible extension to this model is the *integral rectangular parallelepiped (IRPP) model* [20], in which not only a single SV is defined but also a collection of multiple SVs is defined. The IRPP method is widely used to predict SEE rates in the radiation effects community, and it is the standard method specified by the European Cooperation for Space Standardization (ECSS) [21].

3.2.2 *Drift-Diffusion-Collection Criterion*

Alternatively, the *drift-diffusion-collection model* [22] has been proposed to address the limitations observed in the previous approaches. In particular, there is no time dependence in the RPP and IRPP models, and it is thus not possible to study the transient currents that are generated during the ionization of the silicon.

By considering the transport of electron-hole pairs in the semiconductor, it is, however, possible to determine the density of charges that will be collected at the different nodes of the circuit. Consequently, the transient currents can be evaluated as well as their effects on the circuit. With this approach, it is possible to account for the multiple-node charge collection and emerging effects observed in advanced technology nodes.

There are basically three physical mechanisms at play, which are (1) the charges drift due to the electric field, (2) the charges diffuse due to the gradient of charge carrier concentration, and (3) the charge collection at the electrodes of the device. The transport of charges is somehow complex to calculate, and TCAD simulations are generally the most accurate tools for this purpose. However, calculating an SEE cross section requires to use Monte Carlo approach that simulates millions of particles as the use of conventional TCAD tools is not the best solution in terms of calculation time. Alternately, [22] claims that because the electric field is mostly weak in the device, the electrons and the holes are essentially diffusing together (ambipolar diffusion) and are separated at the electrodes. Neglecting interface effects, the track of the ion can then be divided into small elements that spherically diffuse the charges. Each electrode can be divided into small parts that collect the charges. The transient current I_D of each collecting drain node is obtained following the Eq. 3.3 [23]:

$$I_D(t) = q.v \iiint \mathrm{LET}(l) \frac{e^{-\frac{r^2}{4Dt}}}{(4\pi Dt)^{\frac{3}{2}}} \mathrm{dxdydl} \qquad (3.3)$$

where q is the elementary charge, v is the carrier velocity in the junction, $\mathrm{LET}(l)$ is the ion linear energy transfer (LET) along the ion track, r is the distance between the elemental section of the collecting area and the ion track, and D is the ambipolar diffusion coefficient.

For a given ion, at a given location and in a given direction, it is possible to estimate the transient current at each electrode of the device. Finally, based on these currents, it is possible to state whether an SEE occurs or not by using one of the following conditions:

- The total collected charge is higher than the critical charge representative of the device.
- The amplitude of one transient current is high enough and its duration is long enough. The thresholds depend on the device.
- The transient currents are injected in the circuit using simulation program with integrated circuit emphasis (SPICE) simulator and an upset is observed. This approach is the most realistic but requires the knowledge of SPICE model, which is not always easy to get. In addition, running SPICE for each ion track is CPU time-consuming.

3.3   Modeling Radiation-Induced Currents in Circuit Simulations

Circuit simulators, such as simulation program with integrated circuit emphasis (SPICE), are software tools used to solve a system of equations that fundamentally describe the basic functionality of elementary electronic components (transistors, capacitors, resistors, and diodes) within a circuit. Using the drift-diffusion-collection approach allows us to evaluate the shape and amplitude of the transient current at each electrode. These transients will depend on the nature of the ion, its LET, its location, and its direction. When studying single-event effects at the circuit level, the transient current is modeled as a parasitic current source connected to the node of the circuit where the particle hit. It is sometimes useful to use an analytical expression of the transient current, and the analytical model proposed by Messenger [24] is widely used to describe the radiation-induced current as a double exponential transient pulse:

$$I(t) = \frac{Q_C}{t_f - t_r} \cdot \left(e^{-\frac{t}{t_f}} - e^{-\frac{t}{t_r}} \right) \qquad (3.4)$$

where t_r and t_f are the rising and falling time constants, respectively. These time constants correlate to the ionization track formation and the collection efficiency of the p-n junction of the injecting node [25]. Besides being technology dependent, these variables are also influenced by the relative distance of the particle hit. However, several estimation approaches have been proposed based on usually known technology parameters [24–27]. For example, in [27], the double exponential law was verified against the circuit simulation using the transient pulse obtained from a physical model based on the diffusion of carriers. The authors conclude that the double exponential law approach can induce an overestimation of approximately 10–20% on the Single-Event Upset (SEU) cross section of SRAM cells. Additionally, it was shown that the rising time constant t_r can be simplified in the Messenger's analytical model by considering that its value is one fifth of t_f value:

$$I(t) = \frac{5Q_C}{4t_f} \cdot \left(e^{-\frac{t}{t_f}} - e^{-\frac{5t}{t_f}} \right) \tag{3.5}$$

And as proposed by Messenger in [24], the t_f can be calculated based on the minority carrier mobility μ and the substrate doping density of the p-n junction, with the following equation:

$$t_f = \frac{k\epsilon_0\epsilon_{Si}}{q\mu N} \tag{3.6}$$

where k is the Boltzmann constant, ϵ_0 is the permittivity of vaccum, ϵ_{Si} is the relative permittivity of silicon, and q is the electron charge. Once these double exponential law parameters are defined, the radiation-induced current can be injected in the sensitive nodes of the circuit in a transient analysis simulation using SPICE. In Fig. 3.2, the schematic of an inverter and the corresponding current source for each type of particle hit is shown. If the particle strikes an N-channel metal-oxide semiconductor (NMOS) device, the current source is introduced between the drain of the transistor and the ground supply. Conversely, if the particle impacts a P-channel metal-oxide semiconductor (PMOS) device, the current source is connected between the drain of the transistor and the power supply. This distinction arises due to the different operating principles of NMOS and PMOS transistors, leading to different transient behaviors depending on the type of transistor affected by the radiation event.

As discussed in the previous chapters, the charge sharing mechanism in deeply scaled technologies leads to the **multiple-node collection**. Therefore, to take it into consideration the electrical simulations, multiple current sources should be used, increasing the complexity of this methodology.

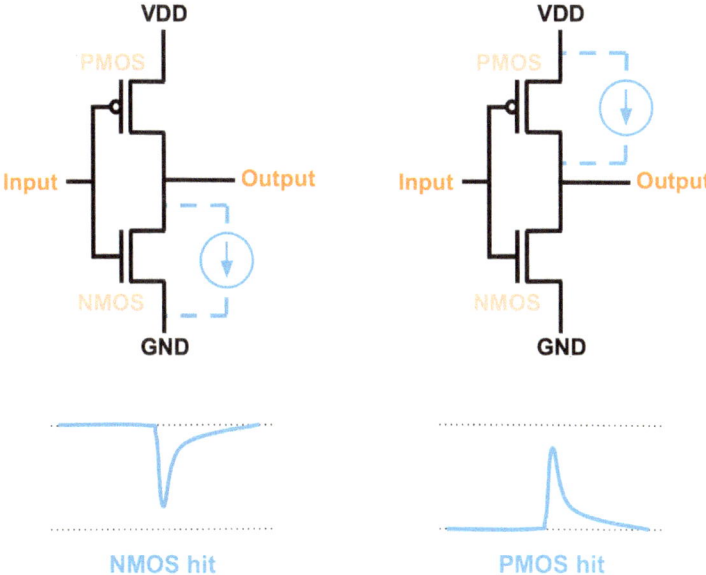

Fig. 3.2 Transient current injection at the circuit level considering a particle hit in NMOS or PMOS devices

3.4   Proposed Prediction Methodology Based on the Diffusion Model

To ensure an accurate assessment of single-event effect (SEE) immunity in digital circuits, it is highly recommended to adopt a multi-scale and multi-physics methodology that considers the complex effects at both the silicon and circuit levels [5, 14]. Various approaches have been developed, encompassing aspects ranging from particle interaction physics to circuit layout design, as discussed in [10]. As aforementioned, emerging effects such as parasitic bipolar amplification (PBA) and charge sharing effects need to be carefully addressed when analyzing radiation interaction in highly scaled technologies [7, 28, 29]. Therefore, layout information from the circuit design is an important determinant of the SEE prediction of electronic circuits. Accordingly, in this work, a layout-based methodology to assess the SEE robustness of digital circuits using the MC-Oracle prediction tool [6] is proposed. MC-Oracle is a Monte Carlo simulation code developed to analyze the SEE immunity of electronics based on the particle interaction physics within the sensitive devices. As neutrons, protons, and ions can be simulated, the circuit sensitivity can be calculated for different radiation environments such as space, atmosphere, ground, and accelerators.

As explained in Chap. 1, energetic particles when interacting with silicon go through energy loss mechanisms such as the ionization process (i.e., generation of electron-hole pairs). This energy loss is responsible for the deposition of a parasitic charge that can be collected by the sensitive transistor junctions and disturb the correct functionality of the circuit. Since neutrons are uncharged particles, they do not experience coulombic interactions with orbital electrons. Consequently, neutrons cannot ionize matter directly, howsoever, it is still considered a threat to electronics in space and aviation applications due to their indirect ionization capability [30, 31]. Considering neutrons can experience nuclear reactions with the material target nuclei, they can induce SEE through the ionization of secondary products of nuclear reactions. Also, as it presents no electromagnetic interaction, neutrons are highly penetrating particles. In the MC-Oracle, the ionization process is modeled using tables of range and electronic stopping power pre-calculated with the stopping and range of ions in matter (SRIM) code [32]. For the nuclear reactions induced by protons or neutrons, a pre-calculated nuclear database for a given energy range is built based on the detailed history of recoiling ions induced by nucleons (DHORIN) code [33]. The location of each nuclear reaction is determined considering the information from the nuclear database in which the mean free path of each particle, i.e., the average distance traveled between collisions is estimated from the nuclear cross section.

In the MC-Oracle simulation, the energy deposition resulting from ionization and nuclear reactions is modeled, followed by the simulation of charge transport and collection using the drift-diffusion mechanism. Hundreds of thousands of particle interactions are simulated, and the resulting paths of ionizing electrons and holes are numerically divided into small fragments to calculate the transport of carriers [6, 34]. To illustrate the layout-based analysis using MC-Oracle, a simplified representation is provided in Fig. 3.3.

Given a graphical design system (GDS) file of the circuit, the collecting drain area of transistors can be identified and extracted to be submitted as input to the MC-Oracle calculations. In this example, the layout design of an inverter logic gate is

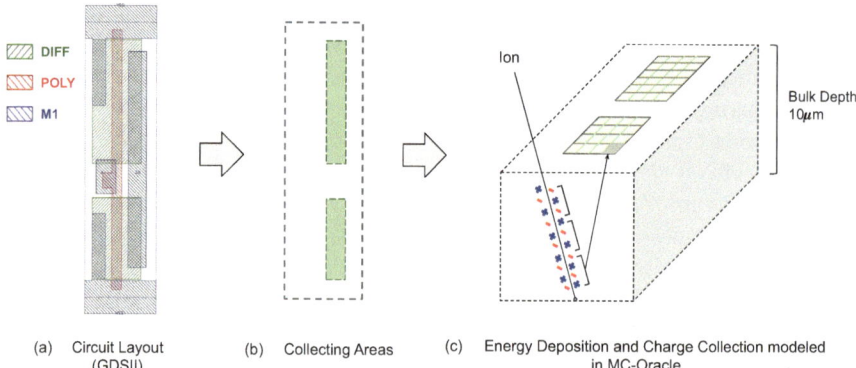

(a)   Circuit Layout   (b)   Collecting Areas   (c)   Energy Deposition and Charge Collection modeled
      (GDSII)   in MC-Oracle

Fig. 3.3 Representation of the extraction of the collecting areas from the circuit layout design (GDSII file) and the energy deposition and charge collection calculation in the MC-Oracle tool. Only the active diffusion (DIFF), poly-silicon (POLY), and metal 1 (M1) layers are shown in the circuit layout for the sake of simplicity

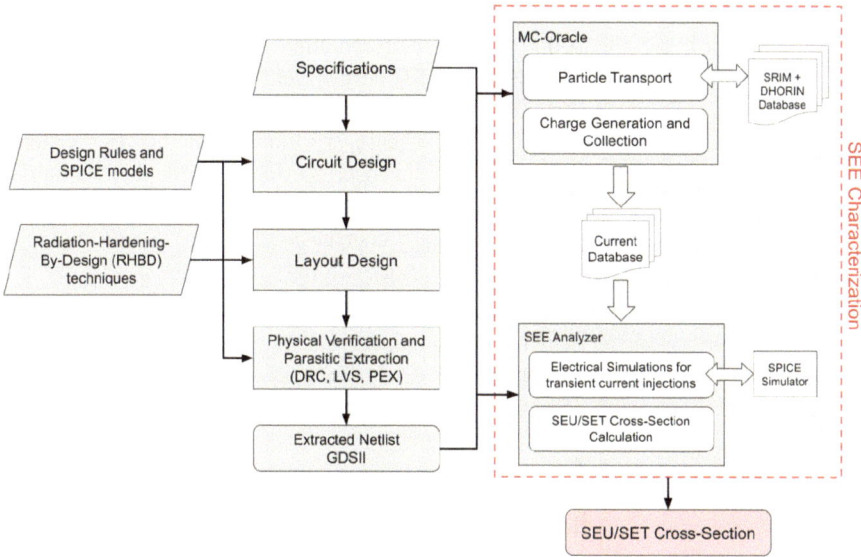

Fig. 3.4 Simulation chain proposed as the SEE prediction methodology

shown in Fig. 3.3a. The drain area of the PMOS and NMOS devices are extracted as shown in Fig. 3.3b. Then, the drift-diffusion-collection model is applied for each ion track of the event (in the case of neutron or proton interaction, one may have several ion tracks simultaneously). Considering an event with a single ion track for the sake of simplicity, Fig. 3.3c shows that the ion track is numerically divided into small fragments in which the generated charges diffuse to the collecting drain areas. Each collecting area is divided into elementary collecting areas and the induced transient current is calculated from the integration of the collected charge along the ionizing track for each elemental section of the collecting area, Fig. 3.3c. For each particle event, MC-Oracle calculates the induced transient current for each collecting area of the circuit design and stores this information in a SET current database. Therefore, multiple-node charge collection effects such as charge sharing mechanism and pulse quenching effects can be evaluated using this tool [28]. A simplified full custom design flow with the SEE characterization methodology using MC-Oracle is shown in Fig. 3.4. Given the specifications concerning the system functionality and reliability (including the radiation environment), the design engineer can start the circuit design process. Once the physical verification, i.e., design rule check (DRC), and layout versus schematic (LVS) are performed, the parasitic extraction of the netlist description and GSDII file can be obtained and submitted to the SET characterization.

The proposed SET characterization is divided into main two steps: first, aiming to build a SET current database, the MC-Oracle tool is used to perform the simulation of the particle transport and the charge collection in the collecting areas of the circuit; second, a SET analyzer is responsible for the SPICE injection campaign using the current database provided by MC-Oracle. The main inputs to the SET characterization are the technology model, radiation environment specification, layout design (GSDII), and extracted netlist description of the circuit. For the SET cross section and pulse width measurement, only the transient pulses with voltage peaks higher than half of the supply voltage are considered, but this criterion can be easily adjusted to the needs of the user. Different hardening techniques can be adopted to prevent the critical electronic systems, such as spacecrafts and avionic control systems, fail due to the occurrence of SEEs. Accordingly, the proposed predictive SET characterization methodology allows the investigation of the hardening effectiveness of radiation hardening by design (RHBD) techniques at the layout level and circuit level.

3.5 Summary

In this chapter, we provided a brief overview of the modeling and prediction of single-event effects (SEEs) in electronic circuits. The increasing computational power and availability of advanced particle physics models have led to the growing use of Monte Carlo-based software tools for studying and predicting the radiation sensitivity of circuit designs.

One commonly used approach is the rectangular parallelepiped (RPP) criterion, which makes assumptions about the sensitive volume of a device. While it has limitations, the RPP approach can provide an order of magnitude estimate of the SEE cross section and allow for the investigation of various parameters such as materials and geometry. For advanced technologies, a more accurate approach is the time-dependent drift-diffusion-collection model, which takes into account charge sharing mechanisms. In electrical simulations using tools like SPICE, the double exponential law is often used to approximate the transient pulse shape, showing reasonably good agreement with the diffusion model, but with a tendency to overestimate cross sections by 10–20%.

We also described the multi-scale and multi-physics simulation chain used in this book as the methodology for SEE prediction. This approach combines particle physics simulations from the MC-Oracle tool with electrical simulations from a SEE analyzer, providing valuable information for the characterization of SEE effects in the circuit designs studied here. Given the importance of layout design in considering SEEs and implementing radiation hardening techniques, our methodology takes into account not only the circuit description in netlist format but also layout design information obtained from the GSDII file. In Chap. 4, we will present and discuss radiation hardening techniques used to enhance the reliability of circuit designs.

Highlights

- Simulation provides a means of testing hypothetical devices or even conditions that are not feasible to be reproduced or measured experimentally.
- Circuit modeling and simulation can support the understanding of several mechanisms before and after an irradiation campaign.
- The integral Rectangular Parallelepiped (IRPP) model is the standard method for the SEE error calculation specified by the European Cooperation for Space Standardization (ECSS).
- Diffusion-collection model is a more accurate alternative to the traditional RPP models, as it takes into account the charge sharing effects.
- SPICE simulations can be used to investigate circuit-level effects.
- The double exponential law proposed by Messenger shows a fairly good agreement with physical models such as the diffusion-collection model.

References

1. Bing J Sheu, Donald L Scharfetter, P-K Ko, and M-C Jeng. BSIM: Berkeley short-channel IGFET model for MOS transistors. *IEEE Journal of Solid-State Circuits*, 22 (4): 558–566, 1987.
2. K-Y Toh, P-K Ko, and Robert G Meyer. An engineering model for short-channel MOS devices. *IEEE Journal of solid-state circuits*, 23 (4): 950–958, 1988.
3. D Munteanu and J-L Autran. Modeling and simulation of single-event effects in digital devices and ICS. *IEEE Transactions on Nuclear science*, 55 (4): 1854–1878, 2008.
4. K. M. Warren, B. D. Sierawski, R. A. Reed, R. A. Weller, C. Carmichael, A. Lesea, M. H. Mendenhall, P. E. Dodd, R. D. Schrimpf, L. W. Massengill, T. Hoang, H. Wan, J. L. De Jong, R. Padovani, and J. J. Fabula. Monte-Carlo based on-orbit single event upset rate prediction for a radiation hardened by design latch. *IEEE Transactions on Nuclear Science*, 54 (6): 2419–2425, Dec 2007. ISSN 0018-9499. doi: https://doi.org/10.1109/TNS.2007.907678.
5. G. Hubert, S. Duzellier, C. Inguimbert, C. Boatella-Polo, F. Bezerra, and R. Ecoffet. Operational SER calculations on the SAC-C orbit using the multi-scales single event phenomena predictive platform (MUSCA SEP^3). *IEEE Transactions on Nuclear Science*, 56 (6): 3032–3042, Dec 2009. ISSN 0018-9499. doi: https://doi.org/10.1109/TNS.2009.2034148.
6. Frédéric Wrobel and Frédéric Saigné. MC-ORACLE: A tool for predicting soft error rate. *Computer Physics Communications*, 182 (2): 317–321, 2011.
7. L Artola, M Gaillardin, G Hubert, M Raine, and P Paillet. Modeling single event transients in advanced devices and ICS. *IEEE Transactions on Nuclear Science*, 62 (4): 1528–1539, 2015.
8. Ygor Q. Aguiar, Laurent Artola, Guillaume Hubert, Cristina Meinhardt, Fernanda Kastensmidt, and Ricardo Reis. Evaluation of radiation-induced soft error in majority voters designed in 7 nm FinFET technology. *Microelectronics Reliability*, 2017. doi: https://doi.org/10.1016/j.microrel.2017.06.077.
9. Steve Koontz, Brandon Reddell, and Paul Boeder. Calculating spacecraft single event environments with FLUKA: Investigating the effects of spacecraft material atomic number on secondary particle showers, nuclear reactions, and linear energy transfer (LET) spectra, internal to spacecraft avionics materials, at high shielding mass. In *2011 IEEE Radiation Effects Data Workshop*, pages 1–8. IEEE, 2011.

10. R. A. Reed, R. A. Weller, A. Akkerman, J. Barak, W. Culpepper, S. Duzellier, C. Foster, M. Gaillardin, G. Hubert, T. Jordan, I. Jun, S. Koontz, F. Lei, P. McNulty, M. H. Mendenhall, M. Murat, P. Nieminen, P. O'Neill, M. Raine, B. Reddell, F. Saigné, G. Santin, L. Sihver, H. H. K. Tang, P. R. Truscott, and F. Wrobel. Anthology of the development of radiation transport tools as applied to single event effects. *IEEE Transactions on Nuclear Science*, 60 (3): 1876–1911, June 2013. ISSN 0018-9499. doi: https://doi.org/10.1109/TNS.2013.2262101.
11. Walter Calienes, Ygor Q. Aguiar, Cristina Meinhardt, Andrei Vladmirescu, and Ricardo Reis. Evaluation of heavy-ion impact in bulk and FDSOI devices under ZTC condition. *Microelectronics Reliability*, 2017. doi:10.1016/j.microrel.2017.06.063.
12. G Hubert, J-M Palau, K Castellani-Coulie, M-C Calvet, and S Fourtine. Detailed analysis of secondary ions' effect for the calculation of neutron-induced SER in SRAMs. *IEEE Transactions on Nuclear Science*, 48 (6): 1953–1959, 2001.
13. Henry HK Tang and Ethan H Cannon. SEMM-2: A modeling system for single event upset analysis. *IEEE Transactions on Nuclear Science*, 51 (6): 3342–3348, 2004.
14. Kevin M Warren, Andrew L Sternberg, Robert A Weller, Mark P Baze, Lloyd W Massengill, Robert A Reed, Marcus H Mendenhall, and Ronald D Schrimpf. Integrating circuit level simulation and Monte-Carlo radiation transport code for single event upset analysis in SEU hardened circuitry. *IEEE Transactions on Nuclear Science*, 55 (6): 2886–2894, 2008.
15. JL Autran, S Semikh, D Munteanu, S Serre, G Gasiot, and P Roche. Soft-error rate of advanced SRAM memories: Modeling and Monte Carlo simulation, numerical simulation-from theory to industry. *edited by Mykhaylo Andriychuk*, pages 309–336.
16. G Battistoni, S Muraro, PR Sala, F Cerutti, A Ferrari, S Roesler, A Fasso, and J Ranft. FLUKA: a multi-particle transport code. In *Proceedings of the hadronic shower simulation workshop*, volume 896, pages 31–49. AIP, 2006.
17. Dávid Lucsányi, Rubén García Alía, Kacper Biłko, Matteo Cecchetto, Salvatore Fiore, and Elisa Pirovano. G4SEE: A geant4-based single event effect simulation toolkit and its validation through monoenergetic neutron measurements. *IEEE Transactions on Nuclear Science*, 69 (3): 273–281, 2022.
18. John N Bradford. Geometric analysis of soft errors and oxide damage produced by heavy cosmic rays and alpha particles. *IEEE Transactions on Nuclear Science*, 27 (1): 941–947, 1980.
19. Robert A Weller, Marcus H Mendenhall, Robert A Reed, Ronald D Schrimpf, Kevin M Warren, Brian D Sierawski, and Lloyd W Massengill. Monte Carlo simulation of single event effects. *IEEE Transactions on Nuclear Science*, 57 (4): 1726–1746, 2010.
20. EL Petersen, JC Pickel, JH Adams, and EC Smith. Rate prediction for single event effects-a critique. *IEEE Transactions on Nuclear Science*, 39 (6): 1577–1599, 1992.
21. ECSS Secretariat. Space engineering: Calculation of radiation and its effects and margin policy handbook - ECSS-E-HB-10-12A. 2010.
22. J-M Palau, R Wrobel, K Castellani-Coulié, M-C Calvet, PE Dodd, and FW Sexton. Monte Carlo exploration of neutron-induced SEU-sensitive volumes in SRAMs. *IEEE Transactions on Nuclear Science*, 49 (6): 3075–3081, 2002.
23. Ygor Q. Aguiar, Frédéric Wrobel, Jean-Luc Autran, Paul Leroux, Frédéric Saigné, Vincent Pouget, and Antoine D Touboul. Mitigation and predictive assessment of set immunity of digital logic circuits for space missions. *Aerospace*, 7 (2): 12, 2020.
24. GC Messenger. Collection of charge on junction nodes from ion tracks. *Nuclear Science, IEEE Transactions on*, 29 (6): 2024–2031, 1982.
25. Paul E Dodd and Lloyd W Massengill. Basic mechanisms and modeling of single-event upset in digital microelectronics. *IEEE Transactions on Nuclear Science*, 50 (3): 583–602, 2003.
26. Marko Andjelkovic, Aleksandar Ilic, Zoran Stamenkovic, Milos Krstic, and Rolf Kraemer. An overview of the modeling and simulation of the single event transients at the circuit level. In *2017 IEEE 30th International Conference on Microelectronics (MIEL)*, pages 35–44. IEEE, 2017.

27. F. Wrobel, L. Dilillo, A. D. Touboul, V. Pouget, and F. Saigné. Determining realistic parameters for the double exponential law that models transient current pulses. *IEEE Transactions on Nuclear Science*, 61 (4): 1813–1818, Aug 2014. ISSN 0018-9499. doi:10.1109/TNS.2014.2299762.

28. Ygor Q. Aguiar, Frédéric Wrobel, J-L Autran, Paul Leroux, Frédéric Saigné, Antoine D Touboul, and Vincent Pouget. Analysis of the charge sharing effect in the set sensitivity of bulk 45 nm standard cell layouts under heavy ions. *Microelectronics Reliability*, 88: 920–924, 2018.

29. RB Schvittz, YQ Aguiar, F Wrobel, J-L Autran, LS Rosa Jr, and PF Butzen. Comparing analytical and Monte-Carlo-based simulation methods for logic gates set sensitivity evaluation. *Microelectronics Reliability*, 114: 113871, 2020.

30. A Taber and Eugene Normand. Single event upset in avionics. *IEEE Transactions on Nuclear Science*, 40 (2): 120–126, 1993.

31. CA Gossett, BW Hughlock, M Katoozi, GS LaRue, and SA Wender. Single event phenomena in atmospheric neutron environments. *IEEE Transactions on Nuclear Science*, 40 (6): 1845–1852, 1993.

32. James F Ziegler and Jochen P Biersack. The stopping and range of ions in matter. In *Treatise on heavy-ion science*, pages 93–129. Springer, 1985.

33. Frederic Wrobel. Detailed history of recoiling ions induced by nucleons. *Computer Physics Communications*, 178 (2): 88–104, 2008.

34. T Merelle, H Chabane, J-M Palau, K Castellani-Coulie, Frédéric Wrobel, Frédéric Saigné, B Sagnes, J Boch, JR Vaille, G Gasiot, et al. Criterion for SEU occurrence in SRAM deduced from circuit and device simulations in case of neutron-induced SER. *IEEE transactions on nuclear science*, 52 (4): 1148–1155, 2005.

Chapter 4
Radiation Hardening

4.1 Introduction

Due to the extensive usage of electronic systems in harsh environments, considerable research efforts have been dedicated to investigating mitigation techniques against radiation effects. The literature offers a wealth of studies on this topic [1–4]. When components or systems are designed and validated to operate in well-defined radiation environments, they are commonly referred to as "radiation-hardened" or "rad-hard" for short. Radiation hardening strategies can encompass a range of approaches, from modifications in the fabrication process to different circuit design implementations.

The European Cooperation for Space Standardization (ECSS) has proposed a classification of radiation hardening techniques based on the abstraction level in application design, which is depicted in Fig. 4.1. One category of techniques falls under radiation hardening by process (RHBP), which involves modifications in the circuit manufacturing process. Examples of RHBP techniques include variations in doping profiles, substrate technology, and use of different materials. These modifications can have an impact on energy deposition and charge collection processes, as discussed in Chap. 1. However, RHBP techniques are often several generations behind the state-of-the-art CMOS technology, leading to suboptimal performance and higher costs.

On the other hand, radiation hardening by design (RHBD) techniques have demonstrated their effectiveness in providing hardness against radiation effects, particularly in highly integrated technologies [4]. RHBD techniques can be applied at different design levels, ranging from the physical layout level to system-level approaches. At the physical layout level, the objective is to reduce radiation-induced charge collection, while at the system level, the focus is on error masking and prevention of system failures. It is worth noting that certain RHBD techniques discussed in this chapter can be applied across multiple design levels.

© The Author(s) 2025
Y. Quadros dc Aguiar et al., *Single-Event Effects, from Space to Accelerator Environments*, https://doi.org/10.1007/978-3-031-71723-9_4

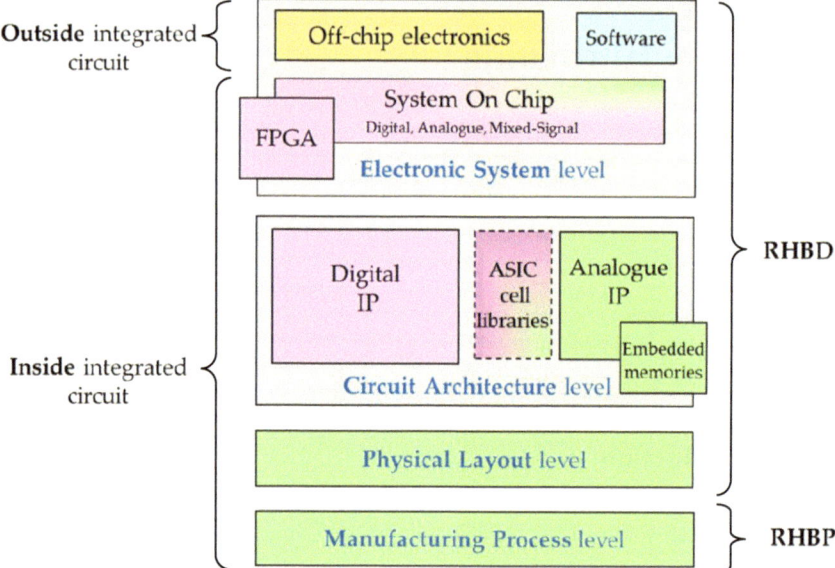

Fig. 4.1 Classification of hardening techniques based on the abstraction level: from manufacturing process to system level (Adapted from [5])

During the design process, the selection of suitable hardening techniques depends on various factors, including the specific radiation environment, the acceptable error rate for the mission, the available system availability, and the design time and resources allocated. The constraints and requirements of each design scenario guide the choice of appropriate hardening techniques. The RHBD techniques can be applied to existing technologies without requiring extensive modifications to the manufacturing process, resulting in faster implementation and reduced costs. Design-level techniques also provide flexibility in tailoring the hardening approach to the specific requirements and constraints of the circuit. This adaptability is crucial as technological advancements continually introduce new challenges.

Over the past decade, there has been a notable shift in the radiation-hardened paradigm from process to design level, propelled by the market in the spacecraft industry known as *New Space*. Historically, government space agencies have played the biggest role in this industry. However, the emergence of New Space has led to the commercialization of space endeavors, resulting in a surge of private companies developing low-cost technologies for space and providing broader accessibility to these technologies. Consequently, a wide range of space-based applications is emerging and increasing the functionality and complexity of space systems. In order to follow this growing market, public and private actors have increased the adoption of the so-called*commercial off the shelf* (COTS) components due to their performance, availability, cost, and lead time. For instance, it is estimated that over 20% of the electrical, electronic, and electromechanical (EEE) components in

European Space Agency (ESA) satellites are COTS [6]. The COTS components are defined as any component designed for commercial purpose only, not following any military or space standard, i.e., not radiation hardened. In this context, design mitigation techniques must be employed to ensure the functionality and performance of the system under radiation effects. This chapter offers a comprehensive review of the foundations and state-of-the-art radiation hardening techniques, focusing specifically on physical layout and circuit architecture.

4.2 Radiation Hardening by Process (RHBP)

Initially, space-qualified components were mostly obtained through the optimization of the manufacturing processes aimed at enhancing the radiation resilience of the process technology itself, i.e., technologies issued from the so-called *rad-hard foundries* [1]. However, these process modifications are often proprietary, making it challenging to access industry-specific information. Nonetheless, these adaptations typically involve the application of different materials, variations in doping profiles, and substrate technologies. For example, studies have demonstrated that removing borophosphosilicate glass (BPSG) layers, commonly used for planarization between metallic layers, can reduce the Single-Event Upset (SEU) rate induced by neutron interactions with boron by about eight to ten times [7, 8]. However, since the adoption of chemical mechanical polishing (CMP) in advanced technologies, BPSG layers are no longer used in the standard manufacturing process, and the main contributor to the thermal neutron SEU rate is the boron isotope ^{10}B present in the source/drain junctions of PMOS devices, p-wells, or tungsten plugs [9]. The natural boron ($_5B$) is abundant in two isotopes: the boron-10 (^{10}B) in 20% and the boron-11 (^{11}B) in 80%. However, the capture cross section of the ^{10}B is three orders of magnitude higher than the ^{11}B, and it is the only one able to release alpha particles that induce Single-Event Effects (SEEs) [10, 11]. Consequently, to mitigate the impact of thermal neutrons, a boron purification process must be integrated into the manufacturing process to reduce the abundance of ^{10}B.

Each additional manufacturing step introduced to the conventional design process adds complexity and increases the fabrication cost. Hence, these rad-hard technologies cannot follow the transistor scaling trend because the high complexity of the manufacturing steps used to achieve radiation hardness leads to higher costs for usually low-volume production. In this way, the available radiation-hardened technology is normally some generations behind the state-of-the-art transistor technology [4]. One example is Sandia's CMOS7 technology process that provides a rad-hard technology based on a 350 nm silicon-on-insulator (SOI) design process [12]; however, the first commercial 350 nm technology process was adopted in mainstream applications in the early 1990s. Thus, besides the higher price, building chips using rad-hard technologies also provides lower performance when compared with highly integrated commercial technologies. Consequently, space systems using rad-hard technology process may face challenges meeting the performance, power,

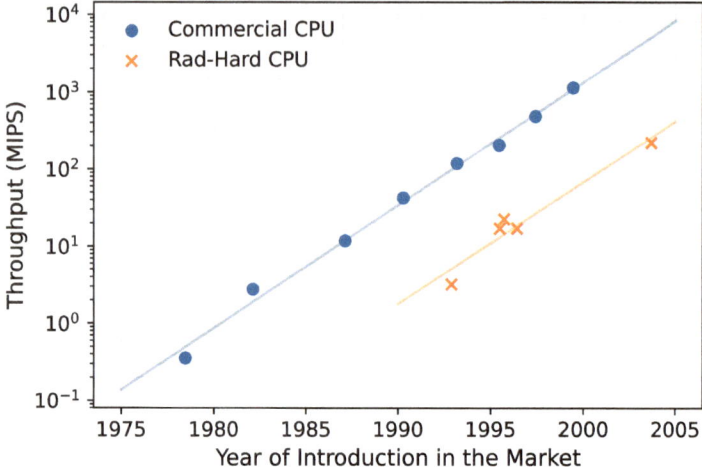

Fig. 4.2 Comparison of commercial and rad-hard processors in terms of throughput (in million instructions per second, MIPS). (Data retrieved from [1])

and area constraints expected in today's space market. This impact can be clearly observed in Fig. 4.2 where the throughput (expressed in million instructions per second, MIPS) is shown for commercial and rad-hard central processing units (CPUs) according to their year of introduction into the market. The throughput of a CPU is an efficiency coefficient in which the number of instructions that a CPU can execute per unit time is estimated for a given clock rate. The rad-hard CPUs providing the same throughput as the commercial CPUs are introduced into the market, in average, 8-10 years after the introduction of the commercial ones [1].

Notes
Radiation hardening, a technology process, increases the fabrication cost and design complexity. Additionally, it is usually limited to technologies generations behind the state-of-the-art ones, leading to lower performances.

One alternative to rad-hard technologies is the adoption of a commercial technology process in which the transistor is built on insulating substrates, i.e., a silicon-on-insulator (SOI) technology [13–15]. Figure 4.3 presents a simplified 2D illustration of an NMOS device fabricated in a bulk technology and in two variants of the SOI technology. In SOI technology, the introduction of an insulation oxide, called a buried silicon oxide (BOX) layer, separates the substrate of the device from its channel and source/drain junctions. In this manner, a reduction of the sensitive volume is obtained, leading to a reduction of the charge collection process in the sensitive nodes and, consequently, improved radiation robustness. Additionally, the

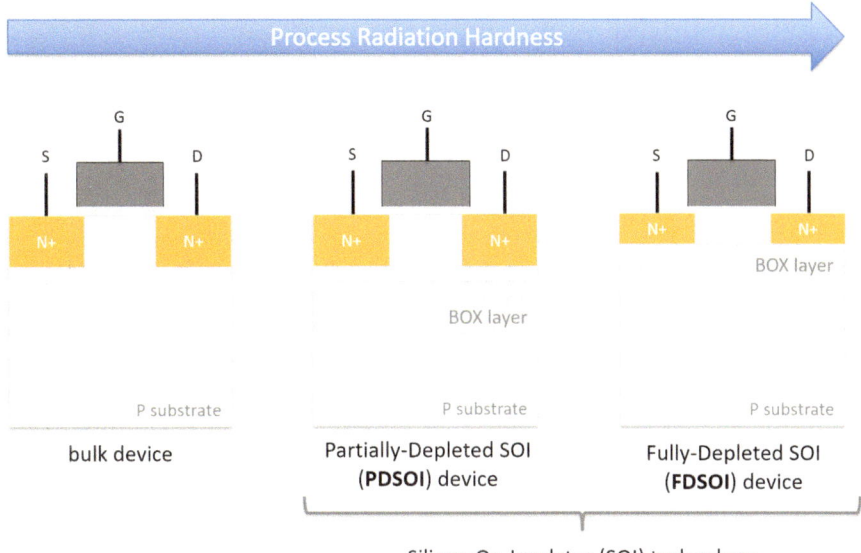

Fig. 4.3 Simplified representation of an NMOS device manufactured in a bulk technology and two variants of a silicon-on-insulator (SOI) technology

BOX layer prevents the charge sharing effect between adjacent devices due to the suppression of the carrier diffusion mechanism. And, very importantly, the SOI device structure eliminates the parasitic silicon-controlled rectifier (SCR) inherently present between transistors in bulk CMOS circuits and is responsible for triggering the latchup mechanism. Thus, SOI-based designs are intrinsically immune to single-event latchup (SEL) effects [13]. However, despite the smaller sensitive volume and immunity to SEL, a stronger *parasitic bipolar amplification (PBA) effect* is observed, and it might degrade the SEE hardness of the SOI-based circuits. When a particle hits the SOI device, the additional carriers can recombine or drift to the p-n junctions. If the majority carriers in the body are able to drift to the source junction and lower the source-to-body potential, an injection of minority carriers from the source can take place, and additional carriers are collected by the drain junction, increasing the magnitude of the SEE, the PBA effect.

Concerning the two SOI variants from Fig. 4.3, the partially depleted SOI (PDSOI) technology presents the closest electrical and structure characteristics to the traditional bulk technology due to the thickness of the silicon film layer on top of the BOX layer. The silicon film layer can be approximately of 50 nm to 200 nm, providing a large and partially depleted body device and high PBA effect. On the other hand, for the fully depleted SOI (FDSOI) technology, the thickness can reach from 5 nm to 20 nm, resulting into a fully depleted body [16]. Due to a thinner silicon film, the FDSOI devices present a higher switching speed and better SEE hardness in response to the stronger charge inversion and low PBA effect. Although there is increase in area, body ties have successfully shown to reduce the bipolar amplification, especially for PDSOI devices [15, 17, 18].

As much as RHBP techniques are quite effective in hardening electronics components against radiation effects, the industry has been increasingly investing on the usage of hardening by design techniques. Specially with the introduction of the New Space market in which the space applications require short entry to the market, lower cost and more onboard processing power, high-level hardening techniques or mitigation strategies at the system level are highly adopted. In the next section, hardening techniques at the layout and circuit levels are presented as they are the main topics discussed in the following chapters.

4.3 Radiation Hardening by Design (RHBD)

4.3.1 Layout-Based Techniques

RHBD techniques can profit from the improvements in power, performance, and reduced area achieved by state-of-the-art commercially available CMOS technology processes [1]. One established RHBD layout technique used to mitigate leakage current induced by total ionizing dose effects is the adoption of edgeless transistors, also known as *enclosed layout transistors (ELT)* [19, 20]. Figure 4.4 presents a layout comparison between the standard and the edgeless transistors. When a

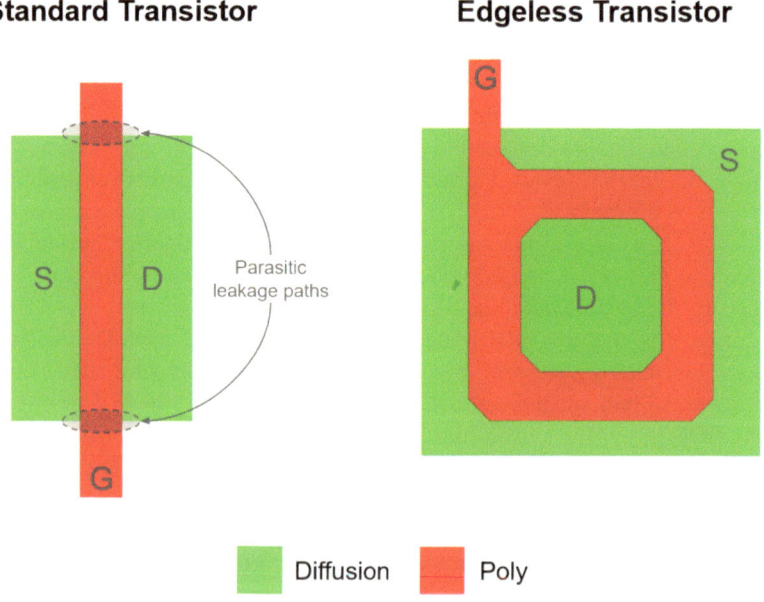

Fig. 4.4 Comparison between a standard transistor layout and the edgeless transistor layout (ELT)

Metal-Oxide Semiconductor (MOS) device is exposed to radiation ionizing dose, positive charges, i.e., holes, can become trapped within the *shallow trench isolation (STI) oxide* near the source and drain junctions. Depending on the density of trapped holes, a parasitic conduction path can be created in the edge of the NMOS transistors [21]. In such cases, inversion in the p-substrate can occur in this parasitic channel formed at the edge between the oxide and the transistor junctions, leading to leakage current flow between the drain and source junctions. The adoption of an enclosed transistor layout helps avoid the connection between the junctions and the sidewall oxides, thereby eliminating the parasitic channel and reducing radiation-induced leakage current [20]. The parasitic leakage paths for the standard transistor layout are indicated in Fig. 4.4.

The biggest drawbacks of this technique are the area overhead and the limitation on the transistor sizing [20, 22]. For instance, the minimum aspect ratio W/L that can be obtained for an edgeless transistor is approximately 2.26. In digital design, density and performance are the priority, thus L is kept the minimum size while W can vary depending on the constraints. However, in analog design, the W/L can be less than 1 as L is increased to reduce leakage currents in low-power designs. Thus, ELT transistors are quite limited when targeting high-performance or low-power designs. Also, due to the layout complexity and the gate geometry, Simulation Program with Integrated Circuit Emphasis (SPICE) models should be adapted to address the nonlinearity of the channel length modulation [23]. Recently, two other layout modifications have been proposed in comparison to the edgeless transistor: the Z-gate [24] and the I-gate transistor layout [25, 26]. Despite the promising results, more studies should be conducted to verify the applicability of these techniques and their consequent drawbacks.

The adoption of edgeless transistors is only capable of preventing leakage current paths within the transistor itself; however, when two NMOS devices are placed side by side a parasitic leakage path can be formed transistor to transistor through the STI oxide. To prevent that, p+ guard rings are used surrounding the NMOS devices, such that the p+ diffusion obstructs any possible parasitic channel between the +n diffusions. Thus, in order to eliminate both intra-device and inter-device radiation-induced leakage paths, *guard-rings* have been vastly used alongside the edgeless transistors [27, 28]. Figure 4.5 illustrates the structure of guard rings around the PMOS device (n+ guard ring connected to power supply voltage) and the NMOS device (p+ guard ring connected to ground supply voltage). As the guard rings provide electrical and spatial isolation, this technique has also shown to provide SEL immunity besides lowering the TID effects [20, 27]. A study was conducted to assess the hardening effectiveness of guard rings schemes against SEL effects in a 180 nm technology [28]. It was shown that both single and dual guard ring configurations (when both PMOS and NMOS devices have guard rings) can provide SEL immunity up to 100 MeV.cm^2/mg. Thus, to lower the area overhead, the single guard ring configuration should be used and prioritizing the p+ guard ring that can also reduce radiation-induced leakage of NMOS devices [28].

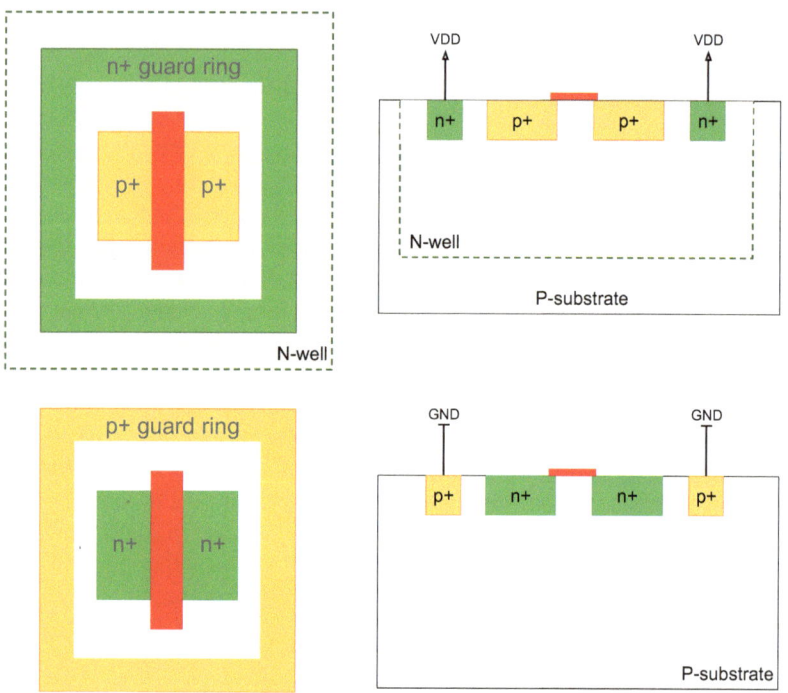

Fig. 4.5 Layout and cross-sectional representation of guard rings around PMOS and NMOS transistors

Notes
Different layout-based techniques can be used alongside in a circuit design; however, their biggest drawbacks are the **increase in area** and **limitation in transistor sizing**.

The transistor positioning within the digital CMOS layout design has also been explored as radiation hardening strategy against SEE [29, 30]. In [29], the placement of NMOS transistors is evaluated in respect to the proximity to the N-well region. Technology Computer Aided Design (TCAD) simulations of an inverter gate with different NMOS positioning ranging from a distance D of 200 nm to 1000 nm from the N-well border are shown in Fig. 4.6. The results show that as closer the device is placed to the N-well region, shorter is the SET pulse width for particle strikes at the NMOS device. This observation was attributed to (1) the reversed-biased diode formed by the N-well and substrate interface which collects the additional carriers and (2) the reinforcement of the recovery current from the PMOS devices while the *off*-state NMOS transistor is hit by the particle. Fig. 4.6 illustrates this phenomenon in which the parasitic bipolar effect in the PMOS device is enhanced by the close

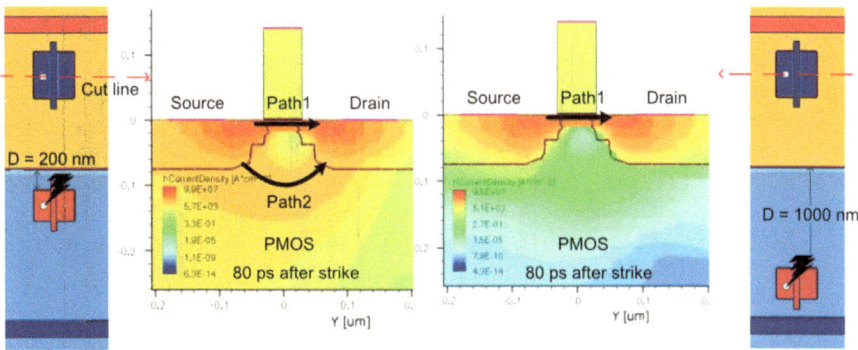

Fig. 4.6 Recovery current reinforcement induced by the increased parasitic bipolar effect due to the close proximity NMOS device. (Adapted from [29])

proximity of the NMOS device. The design with the NMOS device with a distance $D = 200$ nm has nearly ten times the collected charge by N-well in the design with $D = 1000$ nm.

This collected charge is responsible for lowering the electrostatic potential in N-well and activating the parasitic bipolar transistor. With the parasitic bipolar transistor turned *on*, additional carriers will flow from the source to the drain of the PMOS device (Path 2 in the Fig. 4.6), enhancing the recovery current. Although the observed reduction in the SET pulse width, this technique can possibly worsen the SEL resilience of the design, as shown in the TCAD simulations performed in [31]. Due to the activation of the parasitic bipolar transistor, the decrease of the anode-to-cathode (A-C) spacing, i.e., the distance between the PMOS and NMOS devices, has shown a decrease in the threshold LET and an increase in the saturation SEL cross section. Therefore, the designer should have clear in mind the implications of each hardening technique and the target effects intended to mitigate. For instance, the substrate and well tap placement has shown a stronger impact on the SEL sensitivity than the A-C spacing [31]. Thus, to counteract the negative effect of the close proximity NMOS devices, substrate and well tap placement can be used as described in [31].

Another layout technique that uses the transistor placement to improve the SEE robustness is the layout design through error-aware transistor positioning (LEAP) layout technique [30]. In this layout approach, the transistors are placed horizontally in such a way that all the collecting nodes are aligned. The whole idea is to take advantage of the charge collection by the on-state transistors to pull the output voltage back to the expected value. One example of a LEAP-based layout for an inverter gate is shown in Fig. 4.7.

Whenever the particle strikes horizontally, both drain regions will be collecting the additional charges [2, 30]. Due to the horizontal transistor alignment, the charge collected by the *on*-state NMOS transistor will counteract the charge collected by the *off*-state PMOS transistor [30]. Overall, the resulting transient pulse is shortened

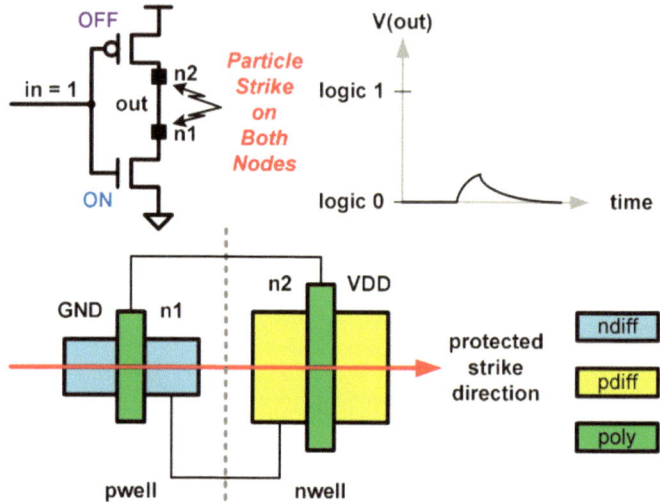

Fig. 4.7 Illustration of the LEAP principle for an inverter gate. When a particle strikes both NMOS and PMOS drain nodes simultaneously, the charge collected by the on-state transistor reduces the overall transient pulse at the output of the gate. (Adapted from [30])

as the *on*-state NMOS is pulling the output voltage back to the logic 0, a similar principle of the recovery current reinforcement from [29].

The transistor sizing (also known as gate sizing) is another well-known hardening technique based on the radiation-induced transient dependency on the drive strength and nodal capacitance of the circuit [32]. However, by increasing the transistor width, the drain area is also increased, and it can induce a greater charge collection process and increased SEE sensitivity. In the next chapter, the applications of gate sizing, transistor stacking, and transistor folding are investigated.

4.3.2 Circuit-Based Techniques

In addition to layout techniques, radiation effects can be mitigated or partially hardened using circuit-based techniques. As circuit complexity increases in advanced technology, susceptibility to physical defects and environmental disturbances, including radiation, necessitates fault tolerance techniques [33–36]. For instance, safety- and mission-critical systems, such as satellites and aircraft flight control systems, heavily rely on fault tolerance methods to enhance reliability. Fault tolerance ensures system functionality remains intact even in the presence of faults, with redundancy forming the cornerstone of many proposed techniques [35]. Unlike hardening techniques aimed at preventing the generation of SEE, i.e., energy deposition and charge collection, fault tolerance focuses on masking the soft errors. Various approaches exist, involving hardware, software, information, and

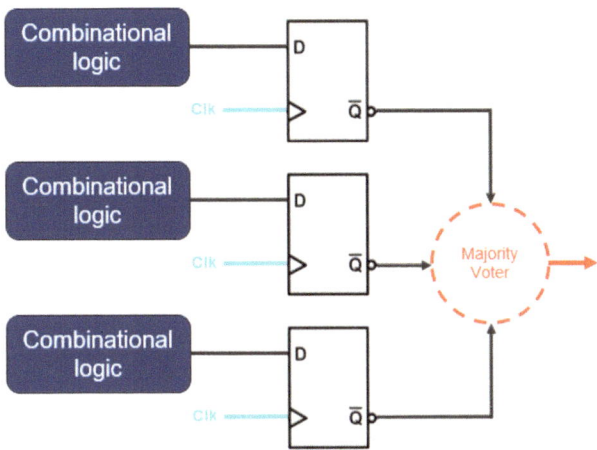

Fig. 4.8 Diagram of the triple modular redundancy (TMR) fault tolerance technique. Combinational and sequential logic are triplicated and output is connected to a majority voter

time redundancy [33]. Hardware redundancy, also known as spatial redundancy, is particularly prevalent in space applications due to its ability to detect and/or correct faults [34]. A notable example is triple modular redundancy (TMR), illustrated in Fig. 4.8, renowned for its effectiveness in providing robustness against SEEs [35].

In this approach, the critical component is triplicated in the design, and all the three output signals are connected to a majority voter (MJV) circuit where the majority of the input signal determines the output signal. In other words, whenever two of the triplicates are fault-free, the correct output will be propagated. The main drawback of this technique is clearly the massive increase in area (two duplicates + MJV circuit) and the consequent increase in power consumption. Also, despite the good fault coverage, the technique relies on the robustness of the MJV circuit, because, even if the three components are fault-free, whenever a particle hits the voter and deposits sufficient charge, the SET or SEU will be propagated. Various TMR variants have been proposed to address these challenges, including diversity TMR (DTMR) [37], where different design implementations are used for triplicated modules, and approximate TMR (ATMR) [38], which employs approximate functions instead of exact replicas of the main function. Both variants offer a balanced compromise between area/performance overhead and fault tolerance against SEEs compared to traditional TMR schemes. For instance, Fig. 4.9 presents an example of a temporal TMR, where delay units are added to filter SET pulses to be latched in all flip-flops. This approach can surely reduce the area overhead as the combinational logic is not triplicated, but can impact the performance of the circuit.

While the redundancy technique was initially largely applied at the circuit level, it has also found application in higher levels such as in field-programmable gate arrays (FPGAs), system on chip (SoC), and software-based redundancy [33, 37, 38].

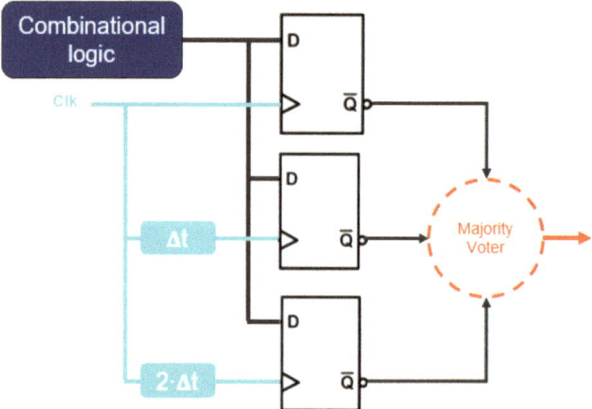

Fig. 4.9 Diagram of the temporal triple modular redundancy (TMR) scheme. Only the sequential logic is triplicated and output is connected to a majority voter, while delay units are used to offset the clock signal and filter the SET pulse from propagating through the majority voter

In contrast to hardware-based TMR implementations, software-level redundancy introduces concerns about execution time overhead.

Redundancy is a mitigation strategy that can be applied in several abstraction levels, from physical and logical to temporal domain.

Another widely used approach at the circuit level is to harden memory cells by using redundant reinforced feedback architectures such as the *dual interlocked storage cell (DICE)* and *Quatro* circuits [39, 40]. These circuit topologies are applied for the design of a 1-bit Static Random Access Memory (SRAM) cell and are shown in Figs. 4.10 and 4.11. Instead of using a single cross-coupled inverter pair as in the standard 6T SRAM design, both circuits have a feedback mechanism sustained by interlocked structures that guarantee the correct output whenever a single sensitive node collects charge. These structures ensure that the stored bit in the memory cell is defined by more than a single pull-down/pull-up network. Although a good radiation response can be achieved, these designs have degraded read/write performance and high area overhead.

The radiation hardness of the DICE cell has been extensively validated through simulations and experiments, being the main reference design for hardening technique in memory circuit designs in the literature [34, 41, 42]. The Quatro design was proposed in [40] as an alternative to the traditional DICE cell with a lower area overhead and better data stability. Some works have shown that flip-flops and SRAM designs based on the Quatro topology have a higher SEU robustness than the DICE cells [43–46]. However, with the close proximity of transistors in deeply

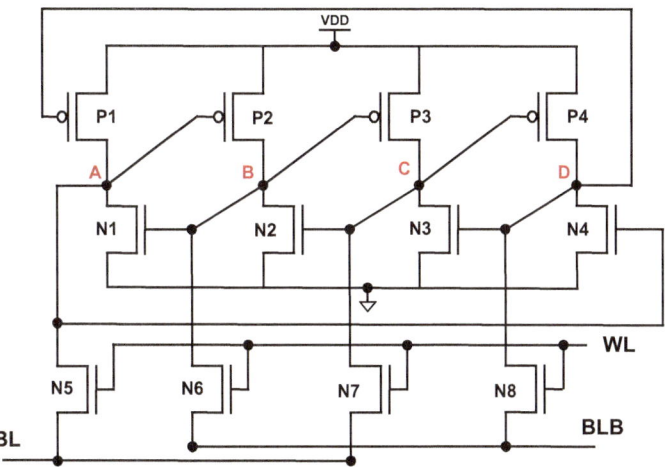

Fig. 4.10 DICE memory cell schematics

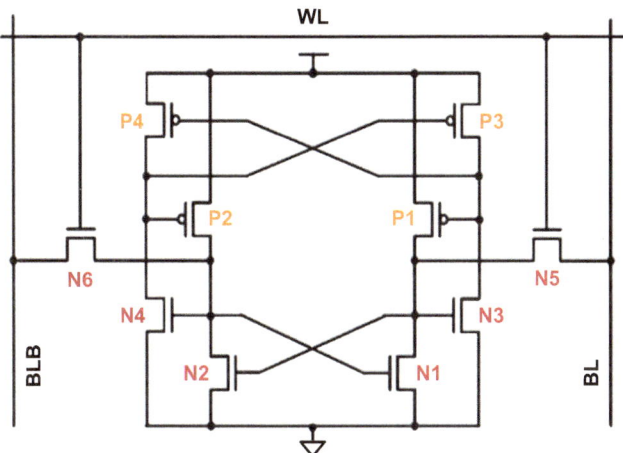

Fig. 4.11 Q10T memory cell schematics

scaled transistor technologies, the impact of charge sharing effects has shown to be a concern for the reliability of these architectures. To address this issue, circuit-based RHBD techniques can be combined along with the layout-based techniques shown in the last section. For instance, the LEAP technique was applied in the DICE design for flip-flops in [2, 30].

The LEAP-DICE flip-flops at 28 nm bulk technology have shown a reduction of the SEU rate of approximately two orders of magnitude when compared to the traditional DICE layout design [2]. Unlike the LEAP technique, which seeks to position specific transistors closer together to leverage charge sharing for improving overall SEE sensitivity, other design approaches focus on increasing nodal spacing.

This strategy is aimed at mitigating the effects of multiple node charge collection in critical nodes. By physically separating these nodes, the probability of a single particle strike causing simultaneous charge collection at multiple sensitive areas is reduced, thereby enhancing the resilience of the circuit to radiation-induced soft errors [42, 47, 48]. With the transistor scaling, the layout design carries a stronger influence on the circuit reliability when considering radiation effects. Accordingly, more and more techniques have been adopted at the layout level to mitigate the effect of multiple node collection and parasitic bipolar amplification [24, 26, 49–53].

4.4 Summary

The increased utilization of electronic systems in radiation environments, including space, aviation, medical applications, and particle accelerators, has underscored the need to explore various hardening techniques to ensure the reliability of these systems. This chapter presented a comprehensive review of fundamental concepts related to radiation hardening techniques against mostly single-event effects (SEE). One approach to enhancing radiation hardness is through process modifications during circuit fabrication, known as radiation hardening by process (RHBP) techniques. These techniques involve altering the manufacturing process to improve radiation resilience. However, with the growing demand for high-performance and low-power solutions, radiation hardening by design (RHBD) techniques have emerged as a promising alternative, particularly when integrated with commercial deeply scaled technologies.

Layout-based RHBD techniques are particularly effective in addressing charge collection mechanisms induced by particle strikes, as they directly impact the charge collection efficiency in sensitive nodes of the circuit. At the circuit level, techniques often utilize redundancy to reinforce logic bits in memory cells or employ voting schemes to mask SEE occurrences. When designing a critical system for a radiation environment, various parameters must be considered to select the appropriate hardening techniques. Combining multiple techniques can achieve the desired level of hardness, but it is important to carefully analyze the associated area, power, and performance overheads to ensure that the reliability requirements are met while respecting design constraints.

The subsequent chapter provides a detailed analysis of layout-based RHBD techniques, offering valuable insights into their implementation and effectiveness. By examining these techniques, designers and researchers can gain a deeper understanding of their capabilities and limitations, enabling them to make informed decisions when developing radiation-hardened systems.

Highlights

- Radiation hardening by process (RHBP) techniques can be used in the manufacturing process to increase the reliability of a transistor technology.
- Radiation hardening by design (RHBD) techniques can be adopted from the physical layout level to the system level to reduce the radiation-induced charge collection or to mitigate system failures.
- Despite the effectiveness of RHBP techniques, their limitations in terms of fabrication cost and performance make the RHBD more attractive for most applications.
- Layout-based techniques can reduce the efficiency of the charge collection process, but one of their limitations is the increase in design area and constraints in the transistor sizing.
- Redundancy is a popularly used mitigation technique that can be implemented in different forms, in the physical, temporal, or information domains, and, therefore, it can be adapted to the most diverse field of applications.

References

1. Ronald C Lacoe. Improving integrated circuit performance through the application of hardness-by-design methodology. *IEEE transactions on Nuclear Science*, 55 (4): 1903–1925, 2008.
2. K Lilja, M Bounasser, S-J Wen, R Wong, J Holst, N Gaspard, S Jagannathan, D Loveless, and B Bhuva. Single-event performance and layout optimization of flip-flops in a 28-nm bulk technology. *IEEE Transactions on Nuclear Science*, 60 (4): 2782–2788, 2013.
3. Wen Zhao, Chaohui He, Wei Chen, Rongmei Chen, Peitian Cong, Fengqi Zhang, Zujun Wang, Chen Shen, Lisang Zheng, Xiaoqiang Guo, et al. Single-event multiple transients in guard-ring hardened inverter chains of different layout designs. *Microelectronics Reliability*, 87: 151–157, 2018.
4. Ronald C Lacoe, Jon V Osborn, Rocky Koga, Stephanie Brown, and Donald C Mayer. Application of hardness-by-design methodology to radiation-tolerant ASIC technologies. *IEEE Transactions on Nuclear Science*, 47 (6): 2334–2341, 2000.
5. ECSS Secretariat. Space product assurance: Techniques for radiation effects mitigation in ASICs and FPGAs handbook. 2016.
6. Mikko Nikulainen. Usage of cots EEE components in ESA space programs, 2019. URL https://escies.org/webdocument/showArticle?id=1064&groupid=6.
7. Robert C Baumann and Eric B Smith. Neutron-induced boron fission as a major source of soft errors in deep submicron SRAM devices. In *2000 IEEE International Reliability Physics Symposium Proceedings. 38th Annual (Cat. No. 00CH37059)*, pages 152–157. IEEE, 2000.
8. Raoul Velazco and Francisco J Franco. Single event effects on digital integrated circuits: Origins and mitigation techniques. In *2007 IEEE International Symposium on Industrial Electronics*, pages 3322–3327. IEEE, 2007.

9. Marcos Olmos, Remi Gaillard, Andreas Van Overberghe, Jerome Beaucour, Shijie Wen, and Sung Chung. Investigation of thermal neutron induced soft error rates in commercial SRAMs with 0.35 μm to 90 nm technologies. In *2006 IEEE International Reliability Physics Symposium Proceedings*, pages 212–216. IEEE, 2006.

10. Elizabeth C Auden, Heather M Quinn, Stephen A Wender, John M O'Donnell, Paul W Lisowski, Jeffrey S George, Ning Xu, Dolores A Black, and Jeffrey D Black. Thermal neutron-induced single-event upsets in microcontrollers containing boron-10. *IEEE Transactions on Nuclear Science*, 67 (1): 29–37, 2019.

11. Daniel Oliveira, Sean Blanchard, Nathan Debardeleben, Fernando F Dos Santos, Gabriel Piscoya Dávila, Philippe Navaux, Carlo Cazzaniga, Christopher Frost, Robert C Baumann, and Paolo Rech. Thermal neutrons: a possible threat for supercomputers and safety critical applications. In *2020 IEEE European Test Symposium (ETS)*, pages 1–6. IEEE, 2020.

12. Kwok-Kee Ma, John Teifel, and Richard S Flores. Sandia rad-hard, fast turn structured asic. Technical report, Sandia National Lab.(SNL-NM), Albuquerque, NM (United States), 2011.

13. O Musseau. Single-event effects in SOI technologies and devices. *IEEE Transactions on Nuclear Science*, 43 (2): 603–613, 1996.

14. JR Schwank, V Ferlet-Cavrois, MR Shaneyfelt, P Paillet, and PE Dodd. Radiation effects in SOI technologies. *IEEE Transactions on nuclear Science*, 50 (3): 522–538, 2003.

15. Philippe Roche, Jean-Luc Autran, Gilles Gasiot, and Daniela Munteanu. Technology downscaling worsening radiation effects in bulk: SOI to the rescue. In *2013 IEEE International Electron Devices Meeting*, pages 31–1. IEEE, 2013.

16. Andrew Marshall and Sreedhar Natarajan. PD-SOI and FD-SOI: a comparison of circuit performance. In *9th International Conference on Electronics, Circuits and Systems*, volume 1, pages 25–28. IEEE, 2002.

17. K Hirose, H Saito, Y Kuroda, S Ishii, Y Fukuoka, and D Takahashi. SEU resistance in advanced SOI-SRAMs fabricated by commercial technology using a rad-hard circuit design. *IEEE Transactions on Nuclear Science*, 49 (6): 2965–2968, 2002.

18. K Hirose, H Saito, S Fukuda, Y Kuroda, S Ishii, D Takahashi, and K Yamamoto. Analysis of body-tie effects on SEU resistance of advanced FD-SOI SRAMs through mixed-mode 3-d simulations. *IEEE transactions on nuclear science*, 51 (6): 3349–3353, 2004.

19. JV Osborn, RC Lacoe, DC Mayer, and G Yabiku. Total dose hardness of three commercial CMOS microelectronics foundries. In *RADECS 97. Fourth European Conference on Radiation and its Effects on Components and Systems (Cat. No. 97TH8294)*, pages 265–270. IEEE, 1997.

20. G Anelli, M Campbell, M Delmastro, F Faccio, S Floria, A Giraldo, E Heijne, P Jarron, K Kloukinas, A Marchioro, et al. Radiation tolerant VLSI circuits in standard deep submicron CMOS technologies for the LHC experiments: practical design aspects. *IEEE Transactions on Nuclear Science*, 46 (6): 1690–1696, 1999.

21. RC Hughes, EP EerNisse, and HJ Stein. Hole transport in MOS oxides. *ITNS*, 22: 2227–2233, 1975.

22. Marc Gaillardin, Martial Martinez, Sylvain Girard, Vincent Goiffon, Philippe Paillet, Jean-Luc Leray, Pierre Magnan, Youcef Ouerdane, Aziz Boukenter, Claude Marcandella, et al. High total ionizing dose and temperature effects on micro-and nano-electronic devices. *IEEE Transactions on Nuclear Science*, 62 (3): 1226–1232, 2015.

23. Varvara Bezhenova and Alicja Michalowska-Forsyth. Aspect ratio of radiation-hardened MOS transistors. *e & i Elektrotechnik und Informationstechnik*, 135 (1): 61–68, 2018.

24. Ying Wang, Chan Shan, Wei Piao, Xing-ji Li, Jian-qun Yang, Fei Cao, and Cheng-hao Yu. 3D numerical simulation of a Z gate layout MOSFET for radiation tolerance. *Micromachines*, 9 (12): 659, 2018.

25. Minwoong Lee, Seongik Cho, Namho Lee, and Jongyeol Kim. Novel logic device for CMOS standard I/O cell with tolerance to total ionizing dose effects. *Solid-State Electronics*, 162: 107630, 2019.

26. Minwoong Lee, Seongik Cho, Namho Lee, and Jongyeol Kim. Design for high reliability of CMOS IC with tolerance on total ionizing dose effect. *IEEE Transactions on Device and Materials Reliability*, 2020.

27. Federico Faccio. Design hardening methodologies for asics. In *Radiation Effects on Embedded Systems*, pages 143–160. Springer, 2007.
28. NA Dodds, NC Hooten, RA Reed, RD Schrimpf, JH Warner, NJ-H Roche, D McMorrow, S-J Wen, R Wong, JF Salzman, et al. Effectiveness of SEL hardening strategies and the latchup domino effect. *IEEE Transactions on Nuclear Science*, 59 (6): 2642–2650, 2012.
29. Zhenyu Wu and Shuming Chen. nMOS transistor location adjustment for N-hit single-event transient mitigation in 65-nm CMOS bulk technology. *IEEE Transactions on Nuclear Science*, 65 (1): 418–425, 2017.
30. Lee Hsiao-Heng Kelin, Lilja Klas, Bounasser Mounaim, Relangi Prasanthi, Ivan R Linscott, Umran S Inan, and Mitra Subhasish. Leap: Layout design through error-aware transistor positioning for soft-error resilient sequential cell design. In *2010 IEEE International Reliability Physics Symposium*, pages 203–212. IEEE, 2010.
31. S Guagliardo, Frédéric Wrobel, Y. Q. Aguiar, J-L Autran, P Leroux, F Saigné, V Pouget, and Antoine Touboul. Single event latchup cross section calculation from TCAD simulations– effects of the doping profiles and anode to cathode spacing. In *IEEE RADECS 2019*, 2019.
32. Quming Zhou and Kartik Mohanram. Gate sizing to radiation harden combinational logic. *IEEE Transactions on Computer-Aided Design of Integrated Circuits and Systems*, 25 (1): 155–166, 2005.
33. Barry Johnson. Fault-tolerant microprocessor-based systems. *IEEE Micro*, (6): 6–21, 1984.
34. Michael Nicolaidis. Design for soft error mitigation. *IEEE Transactions on Device and Materials Reliability*, 5 (3): 405–418, 2005.
35. Fernanda Lima Kastensmidt, Luigi Carro, and Ricardo Augusto da Luz Reis. *Fault-tolerance techniques for SRAM-based FPGAs*, volume 1. Springer, 2006.
36. Robert C Baumann. Radiation-induced soft errors in advanced semiconductor technologies. *Device and Materials Reliability, IEEE Transactions on*, 5 (3): 305–316, 2005.
37. Lucas A Tambara, Fernanda L Kastensmidt, José Rodrigo Azambuja, Eduardo Chielle, Felipe Almeida, Gabriel Nazar, Paolo Rech, Christopher Frost, and Marcelo S Lubaszewski. Evaluating the effectiveness of a diversity TMR scheme under neutrons. In *2013 14th European Conference on Radiation and Its Effects on Components and Systems (RADECS)*, pages 1–5. IEEE, 2013.
38. GS Rodrigues, JS Fonseca, FL Kastensmidt, V Pouget, Alberto Bosio, and S Hamdioui. Approximate TMR based on successive approximation and loop perforation in microprocessors. *Microelectronics Reliability*, 100: 113385, 2019.
39. Teodor Calin, Michael Nicolaidis, and Raoul Velazco. Upset hardened memory design for submicron cmos technology. *IEEE Transactions on nuclear science*, 43 (6): 2874–2878, 1996.
40. Shah M Jahinuzzaman, David J Rennie, and Manoj Sachdev. A soft error tolerant 10T SRAM bit-cell with differential read capability. *IEEE Transactions on Nuclear Science*, 56 (6): 3768–3773, 2009.
41. David G Mavis and Paul H Eaton. Soft error rate mitigation techniques for modern microcircuits. In *2002 IEEE International Reliability Physics Symposium. Proceedings. 40th Annual (Cat. No. 02CH37320)*, pages 216–225. IEEE, 2002.
42. Maxim S Gorbunov, Pavel S Dolotov, Andrey A Antonov, Gennady I Zebrev, Vladimir V Emeliyanov, Anna B Boruzdina, Andrey G Petrov, and Anastasia V Ulanova. Design of 65 nm CMOS SRAM for space applications: A comparative study. *IEEE Transactions on Nuclear Science*, 61 (4): 1575–1582, 2014.
43. S Jagannathan, TD Loveless, BL Bhuva, S-J Wen, R Wong, M Sachdev, D Rennie, and LW Massengill. Single-event tolerant flip-flop design in 40-nm bulk CMOS technology. *IEEE Transactions on Nuclear Science*, 58 (6): 3033–3037, 2011.
44. David J Rennie and Manoj Sachdev. Novel soft error robust flip-flops in 65nm CMOS. *IEEE Transactions on Nuclear Science*, 58 (5): 2470–2476, 2011.
45. Qiong Wu, Yuanqing Li, Li Chen, Anlin He, Gang Guo, Sang H Baeg, Haibin Wang, Rui Liu, Lixiang Li, Shi-Jie Wen, et al. Supply voltage dependence of heavy ion induced sees on 65 nm CMOS bulk SRAMs. *IEEE Transactions on Nuclear Science*, 62 (4): 1898–1904, 2015.

46. Haibin Wang, Jiamin Chu, Jinghe Wei, Junwei Shi, Hongwen Sun, Jianwei Han, and Rong Qian. A single event upset hardened flip-flop design utilizing layout technique. *Microelectronics Reliability*, 102: 113496, 2019.
47. NJ Gaspard, S Jagannathan, ZJ Diggins, MP King, SJ Wen, R Wong, TD Loveless, K Lilja, M Bounasser, T Reece, et al. Technology scaling comparison of flip-flop heavy-ion single-event upset cross sections. *IEEE Transactions on Nuclear Science*, 60 (6): 4368–4373, 2013.
48. Manuel Cabanas-Holmen, Ethan H Cannon, Salim Rabaa, Tony Amort, Jon Ballast, Michael Carson, Duncan Lam, and Roger Brees. Robust SEU mitigation of 32 nm dual redundant flip-flops through interleaving and sensitive node-pair spacing. *IEEE Transactions on Nuclear Science*, 60 (6): 4374–4380, 2013.
49. Jianjun Chen, Shuming Chen, Bin Liang, and Biwei Liu. Simulation study of the layout technique for P-hit single-event transient mitigation via the source isolation. *IEEE Transactions on Device and Materials Reliability*, 12 (2): 501–509, 2012a.
50. Jianjun Chen, Shuming Chen, Yibai He, Junrui Qin, Bin Liang, Biwei Liu, and Pengcheng Huang. Novel layout technique for single-event transient mitigation using dummy transistor. *IEEE Transactions on Device and Materials Reliability*, 13 (1): 177–184, 2012b.
51. Chunhua Qi, Liyi Xiao, Tianqi Wang, Jie Li, and Linzhe Li. A highly reliable memory cell design combined with layout-level approach to tolerant single-event upsets. *IEEE Transactions on Device and Materials Reliability*, 16 (3): 388–395, 2016.
52. Chunyu Peng, Jiati Huang, Changyong Liu, Qiang Zhao, Songsong Xiao, Xiulong Wu, Zhiting Lin, Junning Chen, and Xuan Zeng. Radiation-hardened 14T SRAM bitcell with speed and power optimized for space application. *IEEE Transactions on Very Large Scale Integration (VLSI) Systems*, 27 (2): 407–415, 2018.
53. Jeffrey D Black, Dolores A Black, Nicholas A Domme, Paul E Dodd, Patrick J Griffin, R Nathan Nowlin, James M Trippe, Joseph G Salas, Robert A Reed, Robert A Weller, et al. DFF layout variations in CMOS SOI–analysis of hardening by design options. *IEEE Transactions on Nuclear Science*, 2020.

Chapter 5
Analysis of Layout-Based RHBD Techniques

5.1 Introduction

Radiation robustness can be accomplished through the meticulous application of reliability-aware logic and physical synthesis techniques within semi-custom designs, utilizing the established *standard cell methodology* [1–4].

In essence, this approach enables the hardening of a circuit by selectively employing logic gates that minimize the SET generation or propagation in the most vulnerable nodes of a complex VLSI design. In semiconductor design, the fundamental constituents known as "*standard cells*" serve as the elemental building blocks for the physical realization of Boolean logic functions, including NAND, NOR, inverter, among others. By adopting a cell-based design methodology, the power and performance characteristics of these logic cells become readily accessible, enabling the application of various synthesis algorithms aimed at optimizing these metrics for the entire system design.

A **standard cell library** contains a group of available standard cells that can be used to synthesize a whole system from the RTL description to the final layout physical design.

In [1], three selective node hardening techniques were examined when applied in the logic synthesis of different ISCAS85 benchmark circuits. Remarkably, the findings underscored the remarkable efficacy of these hardening techniques when deployed in standard cell-based VLSI designs. Furthermore, during physical synthesis, the integration of hardening strategies within the cell placement phase proves crucial in mitigating charge sharing effects and promoting pulse quenching effects in electrically interconnected combinational circuits [2–4]. In a notable study

Y. Quadros de Aguiar et al., *Single Event Effects, from Space to Accelerator Environments*, https://doi.org/10.1007/978-3-031-71723-9_5

by Du et al. [3], it was established that as the feature size diminishes, the influence of cell placement on the susceptibility of complex VLSI designs to soft errors becomes more pronounced due to the intricate multi-node collection process. Consequently, there exists an inherent necessity to explore selective node hardening strategies that can seamlessly be integrated into standard cell-based design methodologies. This section offers a comprehensive comparative analysis of the application of gate sizing and transistor stacking techniques within the context of a standard cell methodology, shedding light on their respective merits and limitations.

5.2 Part I—Gate Sizing and Transistor Stacking

5.2.1 Gate Sizing (GS)

The performance of circuits is directly influenced by the feature sizes of transistors, encompassing both the length (L) and width (W) of the device channel. Transistor and gate sizing techniques have gained widespread usage in numerous applications to optimize the trade-off between delay and power consumption [5, 6]. Figure 5.1 provides a simplified illustration of a transistor layout. The diffusion area, represented by the green layer, corresponds to regions where p-n junctions are formed, such as the source (S) and drain (D) of the transistor. The gate (G) of the device is depicted by the red rectangular shape, which represents the polysilicon layer. The blue color represents the metal 1 layer, while the yellow is a via, known as VIA0, that connects metal 1 layer to the diffusion layer.

The electrical characteristics of a transistor device are determined by the width (W) and length (L) of its channel. This is evident in the relationship governing the drain current I_D, as depicted by Eq. 5.1. Altering the W/L ratio of transistors has a significant impact on the nodal capacitance and drive strength of the circuit, thereby influencing power consumption and propagation delay. In digital design, the length (L) of the transistor is typically set to the minimum available value dictated by the technology, as it is a process parameter constraint. Conversely, the width (W) of the transistor can be adjusted to fulfill the system design requirements. Increasing the width (W) reduces resistance, resulting in a higher drain current (I_D) and reduced propagation delay. However, this increase in width also leads to an augmented capacitance, consequently raising power consumption. Thus, the designer must seek the optimal sizing that aligns with the circuit requirements.

$$I_D \propto \frac{W}{L}. \tag{5.1}$$

In addition to their influence on power consumption and propagation delay, the drive strength and nodal capacitance of a circuit also impact the radiation-induced transient currents. Therefore, gate sizing has also been employed to enhance the radiation robustness of VLSI circuits [7]. Let us analyze the radiation response

Transistor layout **Transistor Sizing**

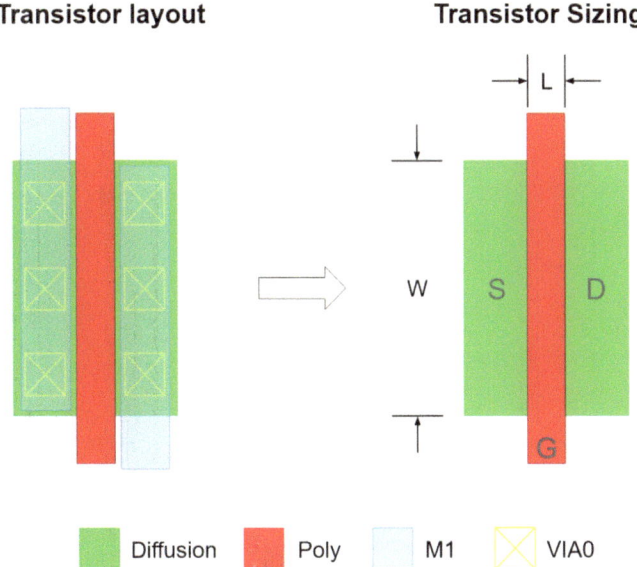

| | Diffusion | | Poly | | M1 | ⊠ | VIA0 |

Fig. 5.1 MOS transistor layout and its main feature sizing: width W and length L of the transistor channel

Table 5.1 Gate sizing scenarios for the SET injection at an inverter

	Wp (nm)	Wn (nm)
Scenario 1	630	415
Scenario 2	945	415
Scenario 3	630	622
Scenario 4	945	622

of an inverter gate across different sizing scenarios (described in Table 5.1). The inverter, designed in the Complementary MOS (CMOS) logic style, consists of a single PMOS transistor in the pull-up network and a single NMOS transistor in the pull-down network. Figure 5.2 illustrates the transistor schematics, truth table, and the gate symbol of a CMOS inverter. The PMOS device is characterized by the channel width denoted as Wp, while the NMOS device is denoted as Wn. When an input signal at the low logic value ("0" or ground) is applied, the output of the inverter assumes a high logic level ("1" or power supply) and vice versa.

Scenario 1 corresponds to the minimum available sizing for an inverter gate in the 45 nm OpenCell NanGate library, a standard cell library for 45 nm technology [8]. Scenarios 2 to 4 involve multiplying the Wp and/or Wn values by a factor of 1.5. To explore the intricate relationship between gate sizing, nodal capacitance, restoring current, and the resulting Single-Event Transient (SET) pulse, electrical simulations are conducted using a SPICE circuit simulator, as described in Chap. 3. A double exponential transient current with consistent arbitrary SET parameters is injected across all sizing scenarios. The goal is to analyze the impact of gate sizing on the

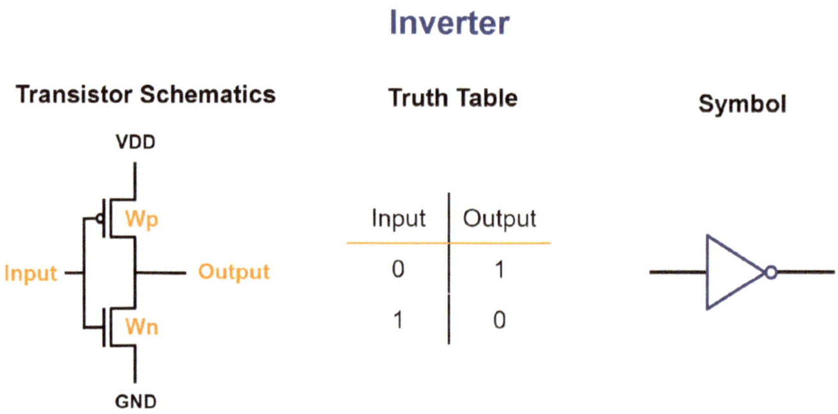

Inverter

Transistor Schematics   **Truth Table**   **Symbol**

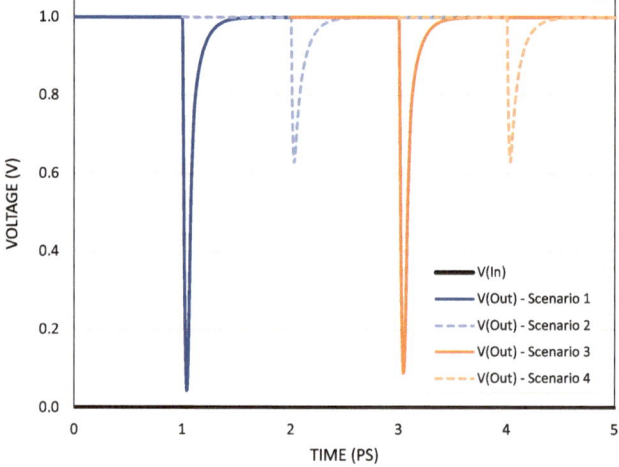

Input	Output
0	1
1	0

Fig. 5.2 Transistor schematics of a CMOS inverter, the truth table, and its symbol

Fig. 5.3 SET injection at the output of a CMOS Inverter with different sizing scenarios when the NMOS device is sensitive

SET response. Figure 5.3 depicts the SET response for the injection campaign, with the input of the gate set to a low logic level. In this configuration, the NMOS device remains in the *off* state, while the PMOS device is responsible for maintaining the output signal at a high logic level. As a reference for this analysis, the SET response of the minimum sizing scenario (Scenario 1) is considered.

In Scenario 2, where only Wp is increased by a factor of 1.5, the resulting SET pulse experiences a significant reduction in duration. However, the peak of the pulse does not reach half of the supply voltage, suggesting that it may be effectively masked in the subsequent gate stage. Conversely, in Scenario 3, when only Wn is upsized, the SET pulse remains nearly unchanged compared to the minimum sizing configuration in Scenario 1. Since the PMOS device is responsible

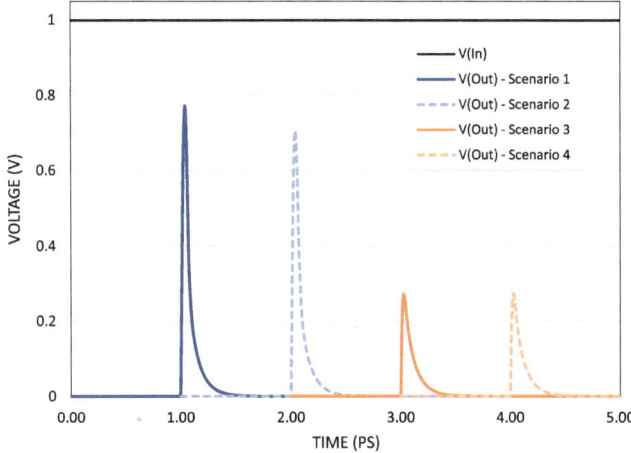

Fig. 5.4 SET injection at the output of a CMOS inverter with different sizing scenarios when the PMOS device is sensitive

for supplying the restoring current, increasing the width of the NMOS transistor (Wn) does not contribute to SET recovery in this case. In this analysis, the same parameters were utilized for defining the current source, resulting in the same amount of collected charge for all sizing scenarios. Nevertheless, by upsizing the transistor, a larger collection area is achieved, potentially enhancing the efficiency of charge collection. Therefore, if the charge collection mechanism were taken into account in this analysis, the resulting SET in Scenario 3 would likely be larger due to the increased collection efficiency facilitated by the augmented width (Wn) of the NMOS transistor. In Scenario 4, where both transistors are upsized, a similar response to Scenario 2 is observed. This outcome further confirms that, when the input signal is at a low logic level, the PMOS device should be upsized to mitigate radiation-induced transient pulses within the inverter gate.

Figure 5.4 presents the same analysis, but with the PMOS device in the *off* state, corresponding to an input signal at a high logic level. Notably, the SET pulse is significantly shortened only when the width of the NMOS transistor (Wn) is increased, as observed in Scenarios 3 and 4. Consequently, Scenario 4 emerges as the optimal sizing choice for enhancing the robustness of the inverter in both input cases. However, it is important to consider that upsizing transistors not only increases capacitance and restoring current but also enlarges the sensitive area. This can potentially compromise the reliability of the circuit by intensifying particle incidence probability and enhancing charge collection efficiency. Standard cell libraries provide cells with varying drive strengths, starting from the minimum-sized implementation (denoted by X1) and incrementing discretely to higher drive strengths like X2, X4, and so on. Due to the regular layout structure of standard cells and the availability of multiple drive strength options, gate sizing within the standard cell methodology is a discrete process. In a study by Cannon et al. [9], an RHBD cell library at 90 nm was evaluated under heavy-ion and high-energy proton irradiation, focusing on inverters, NAND, and NOR logic gates. Different drive strengths offered by the cell library were assessed, revealing that upsizing the cells effectively reduced the SET cross section for the inverter and NOR logic gates. However, in the case of the NAND2_X2 cell, the increased sensitive area dominated the SET sensitivity, outweighing the benefits of increased nodal capacitance and restoring current [9].

> The efficiencies of layout-based techniques are closely dependent on the transistor technology. Therefore, they should be carefully analyzed through layout-based simulations and fault injection electrical characterization.

On the other hand, studies on FinFET-based circuits employing NAND and NOR gates have revealed similar SET sensitivity [10]. The symmetrical sizing of PMOS and NMOS transistors, achieved through strain engineering and width quantization in FinFET technologies, results in comparable drain area and restoring current, thereby leading to equivalent susceptibilities to soft errors for both gate types. In this chapter, gate sizing is evaluated using the prediction methodology elucidated in Sect. 3.4. By employing a comprehensive multi-physics prediction methodology, the complex relationship between charge collection efficiency and electrical characteristics involved in layout-based radiation hardening techniques is examined, providing insights into their impact on the Single-Event Effects (SEE) cross section.

5.2.2 Transistor Stacking (TS)

An alternative approach to gate sizing is the utilization of transistor stacking (TS) to enhance the nodal capacitance [1]. Stacking devices is a well-known RHBD technique used for SEU immunity in SOI designs [11, 12]. To illustrate the principle behind this hardening technique, Fig. 5.5 depicts the transistor schematics of an inverter and a simplified representation of the NMOS transistors at the device level in a SOI technology. Instead of employing two transistors as shown in Fig. 5.2, the TS-based CMOS inverter incorporates four transistors, with the additional transistors connected in series with the original ones.

The use of *shallow trench isolation* (STI) and *buried oxide* (BOX) in SOI structures effectively mitigates charge sharing between stacked transistors, significantly enhancing the overall resilience to soft errors [11]. In a TS inverter employing SOI technology, if an incident particle strikes only transistor N2, the resulting SET is unable to propagate to the output since transistor N1 is in *off* state and behaves as

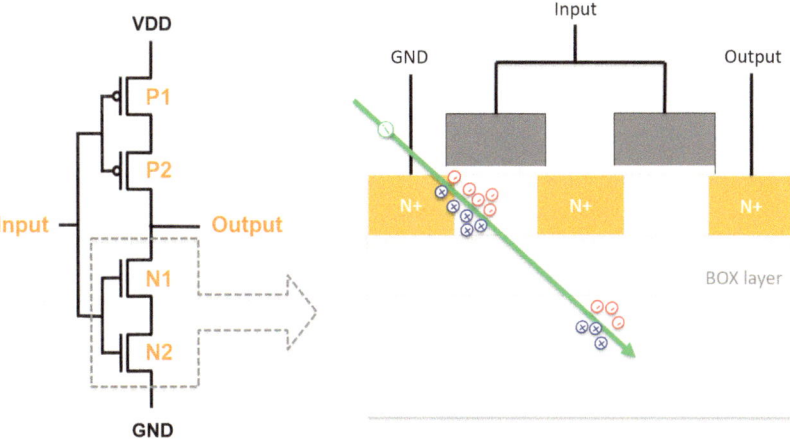

Fig. 5.5 Application of transistor stacking in an inverter design based on a SOI technology and the representation of a heavy-ion ionization. The electron-hole pairs generated within the insulator are omitted due to its negligible contribution to the SEE effects

an open circuit. Hence, for a soft error to manifest in the circuit, a single particle impact must deposit a sufficient charge in both stacked devices, or at least in the nearest one to the output. Recent studies have demonstrated that by exclusively utilizing stacked NMOS devices in SOI latch designs, the SEU rate can be improved by over 80% [13]. In contrast, bulk devices experience charge sharing effects among adjacent transistors. Nonetheless, they still benefit from increased nodal capacitance and the masking effect provided by stacked devices. Moreover, transistor stacking enables power consumption reduction due to lower leakage current compared to single transistors of the same size [14]. In terms of power consumption, transistor stacking has been shown to outperform gate sizing while maintaining similar area efficiency [1]. However, it is important to note that connecting transistors in series, as done in the stacking technique, increases the effective (dis)charging resistance, leading to increased delay and potential degradation of circuit performance.

One of the advantages of gate sizing and transistor stacking is their compatibility with both full-custom designs and standard cell libraries. As explained earlier, a cell library comprises various logic functions implemented in different drive strengths and with varying numbers of input signals. Consequently, a 4-input NAND gate (NAND4) can function as a 2-input NAND gate (NAND2) by incorporating stacked devices, with the additional inputs connected to the original inputs, as illustrated in Fig. 5.6. However, the transistor stacking technique is exclusively applied in the pull-down network, while in the pull-up network, the additional transistors are connected in parallel to enhance the overall driving strength. Similarly, employing a 4-input NOR gate enables the creation of a 2-input NOR gate with PMOS stacked devices.

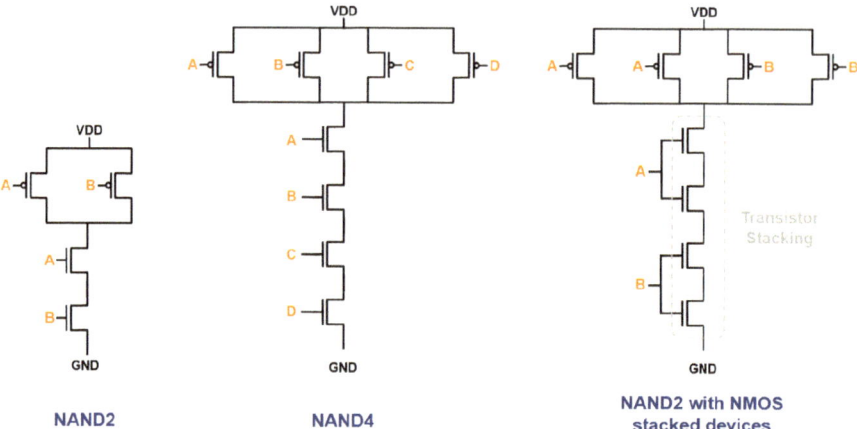

Fig. 5.6 Using 4-input NAND gate to achieve stacking transistors for a 2-input NAND function with standard cell libraries

The following section examines the impact of circuit layout on the deposition and charge collection process, specifically examining the radiation hardening effectiveness of gate sizing and transistor stacking within a cell-based methodology. With a focus on low-power reliable applications, the analysis centers on striking a balance between power consumption and resilience against radiation-induced effects, particularly SETs. As it will be shown, it is of significant importance the exploration of the input dependency associated with each technique, which can be leveraged to enhance the radiation reliability of hardened standard cell libraries while minimizing area, power, and performance overhead.

5.2.3   Comparison of Power and Area Overhead

In terms of power consumption, adopting RHBD techniques often leads to an increase in power, particularly in terms of static power consumption, which is considered in this analysis. Figure 5.7 illustrates the static power consumption for the NAND and NOR gates when employing gate sizing and transistor stacking techniques. For both standard cells, gate sizing demonstrates the highest increase in power consumption, with a factor of 2. This can be attributed to the increased leakage current associated with the adoption of larger transistors.

In contrast, transistor stacking exhibits lower power consumption compared to gate sizing. This is due to the reduction in leakage current resulting from the transistor stacking effect. It is important to note that when adopting transistor stacking using standard cells, the stacked devices for the NAND and NOR gates are the NMOS and PMOS transistors, respectively. Since NMOS transistors in the

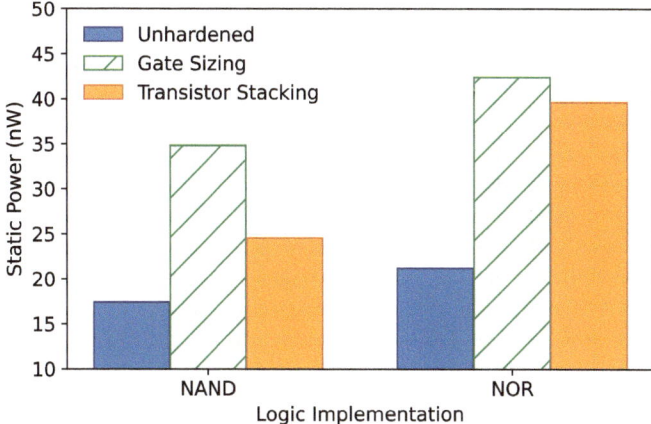

Fig. 5.7 Static power consumption estimation for unhardened, gate sizing, and transistor stacking implementations of the NAND and NOR gates

pull-down network generally have higher leakage current than PMOS transistors, the reduction in static power consumption is more significant for the NAND gate, which incorporates stacked NMOS devices. As a result, the NAND gate consistently displays the lowest static power consumption among the three designs considered in this analysis.

> Layout-based hardening techniques can negatively impact the performance of the circuits, leading to increase in power consumption or propagation delays.

5.2.4   Impact on the SET Cross Section

Figure 5.8 displays the log–log representation of the SET cross-section curves for the NAND gates: unhardened, GS-based, and TS-based design versions. The cross section is calculated for each input signal combination, and the arithmetic mean for each particle LET is presented. At a particle LET of $2.5 \, \text{MeV.cm}^2/\text{mg}$, gate sizing achieves the highest reduction in SET cross section, approximately 78%, while transistor stacking results in a reduction of around 24%. The effectiveness of both techniques in enhancing radiation robustness diminishes as the particle LET increases. Despite the increase in drain area, gate sizing still leads to a reduction in the overall SET cross section, albeit only 3%, for a particle LET of $78 \, \text{MeV.cm}^2/\text{mg}$.

Conversely, the transistor stacking technique increases the SET sensitivity of the circuit by approximately 11.7%. This rise in cross section is attributed to the

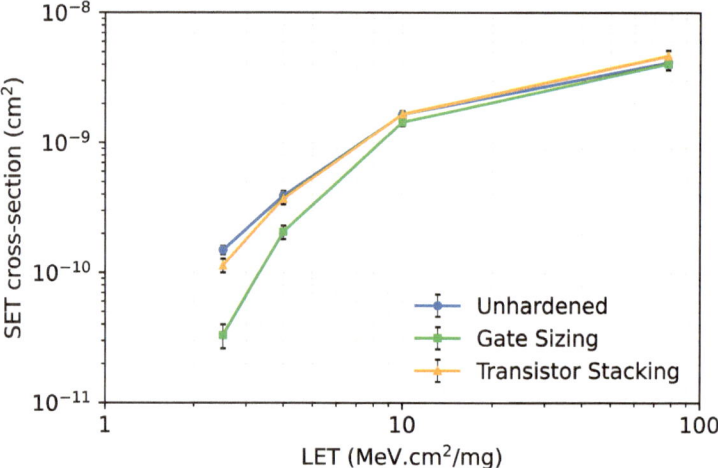

Fig. 5.8 Average of the SET cross section for each input signal the NAND logic gate: minimum sized (unhardened), using gate sizing; and transistor stacking

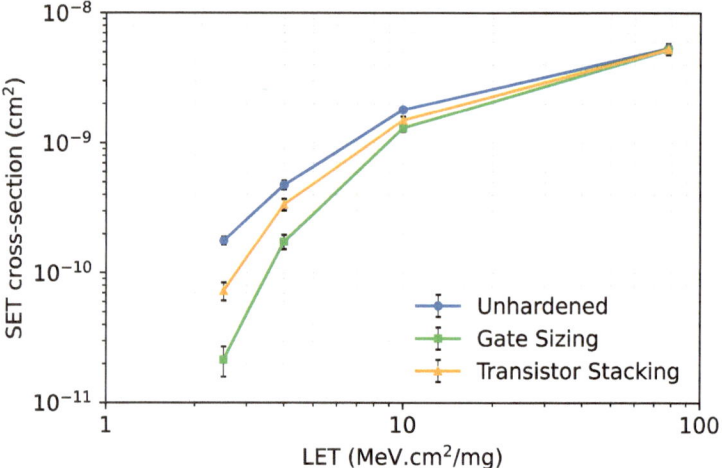

Fig. 5.9 Average of the SET cross section for each input signal the NOR logic gate: minimum sized (unhardened), using gate sizing; and transistor stacking

enlarged layout area and drain regions resulting from these layout-based hardening techniques. For high particle LET, the dominant factor affecting circuit reliability is the enhanced charge collection efficiency facilitated by the larger transistors. Similar trends can be observed in the SET cross-section curves of the NOR gates depicted in Fig. 5.9. However, both techniques exhibit higher efficiency compared to the NAND gates. For instance, transistor stacking achieves an SET cross-section reduction of approximately 60% for the TS-based NOR gate under particles of $2.5\,\mathrm{MeV.cm^2/mg}$,

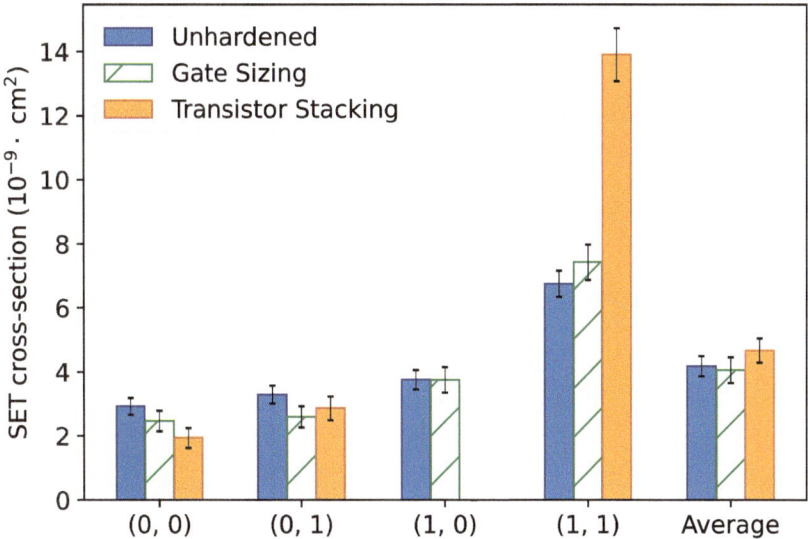

Fig. 5.10 SET cross section for each input signal combination of the NAND gate under 78 MeV.cm^2/mg

twice the reduction observed for the NAND gate. This discrepancy can be attributed to the reduced difference in drain area between the stacked-device NAND and NOR gates and the interplay between the driving capabilities of the pull-up and pull-down transistor networks in the two gates. To gain a deeper understanding of these results, a closer examination of the input signal and layout design of each gate is necessary.

In Fig. 5.10, the SET cross section σ_{SET} for each input signal combination is shown for the NAND gate under a particle LET of 78 MeV.cm^2/mg. The input signal combination (0, 0) exhibits the lowest sensitivity, while the input combination (1, 1) is the most SET sensitive for the NAND gate. When the input signal combination (0, 0) is applied to a NAND gate, both NMOS transistors are turned off, and the two PMOS transistors supply the output signal. This configuration leads to lower sensitivity due to the increased capacitance in the two series-connected NMOS transistors and the strong recovery current from the two parallel-connected PMOS transistors. Conversely, for the NOR gate, which is the complement of the NAND gate, the opposite behavior can be observed in Fig. 5.11. In this case, the input combination (0, 0) is the most SET sensitive, while the input combination (1, 1) is the least sensitive.

Notice that for the most sensitive input combination, both techniques provide a higher cross section than the unhardened design, especially the transistor stacking. In order to explain why the TS technique worsened the reliability in some input scenarios, it is important to analyze the transistor schematics and its equivalent driving strength of each transistor network. In Fig. 5.12, the transistor schematics and the equivalent driving strength of each NAND design are depicted. The

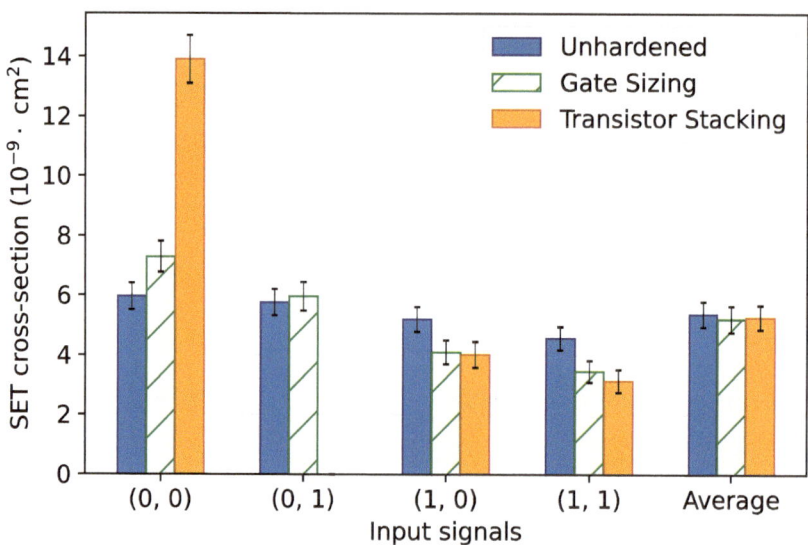

Fig. 5.11 SET cross section for each input signal combination of the NOR gate under 78 MeV.cm^2/mg

unhardened design (NAND2_X1) has its equivalent pull-up and pull-down driving strengths labeled as Wp_{eq} and Wn_{eq}, respectively. Based on the parallel and series associations, an estimation of the driving strength for the GS-based and TS-based designs is provided. As expected, the GS-based design (NAND2_X2) exhibits twice the driving strength of the unhardened design for both the pull-up and pull-down transistor networks. However, in the TS-based design (NAND4_X1), which is logically used as a 2-input NAND gate, only the pull-up network experiences an increase in driving strength, while the pull-down network has its strength reduced by 1.5 times. This explains the observed results for the input combination (1, 1) in Fig. 5.10.

Considering the worst-case input scenario for the transistor-stacked NAND (NOR) design, where all radiation-sensitive transistors are PMOS (NMOS) devices, the 4-stacked NMOS transistors in the NAND pull-down network serve to provide the restoring current to counteract the parasitic SET pulse from the PMOS devices. This configuration helps explain the increased cross section observed for certain input scenarios in Figs. 5.10 and 5.11, particularly when the input combination involves all ones or zeros.

However, transistors in series provide less current drive due to the increased effective resistance, leading to performance degradation and also larger SET pulse width and increased cross section. Additionally, the total drain area of PMOS devices in the TS-based NAND design is twice the area of the unhardened one, inducing a higher collected charge. These can also be confirmed in Fig. 5.13, which the SET pulse width measurements for each technique applied on the NAND gate are provided. The unhardened and gate-sized designs present similar pulse width

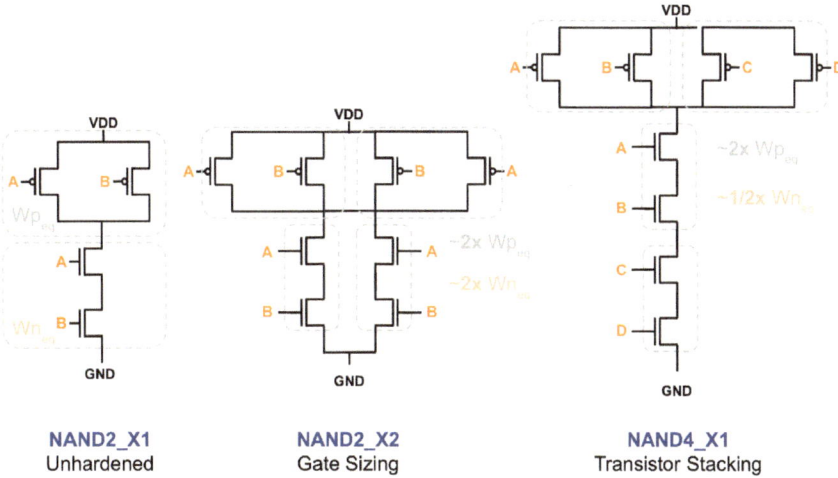

Fig. 5.12 Transistor schematics and the equivalent driving strength of NAND2_X1 (unhardened), NAND2_X2 (gate sizing) and NAND4_X1 (transistor stacking)

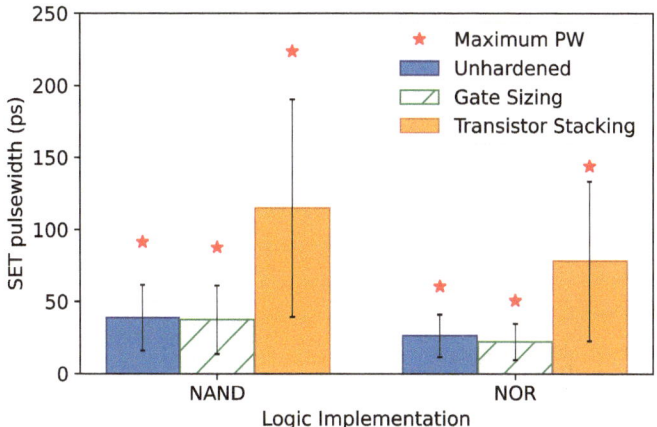

Fig. 5.13 SET pulse width measurements for the NAND and NOR gates under $78\,\text{MeV.cm}^2/\text{mg}$

mean and maximum, while the transistor stacking design can have a maximum SET pulse width more than $2\times$ larger than the unhardened design. As expected from the previous results, the transistor stacking technique increases the overall pulse width mean due to the reduced drive strength of the stacked devices. The similar pulse width features between the unhardened and gate sizing design can be attributed to the balance between the increase of the recovering drain current and the collected charge in the upsized transistors.

Another observation from Figs. 5.10 and 5.11 is that for the TS-based designs, no SET is observed for the input (1, 0) and the input (0, 1) in the NAND and NOR gates, respectively. To understand this result, we need to have a look in the transistor

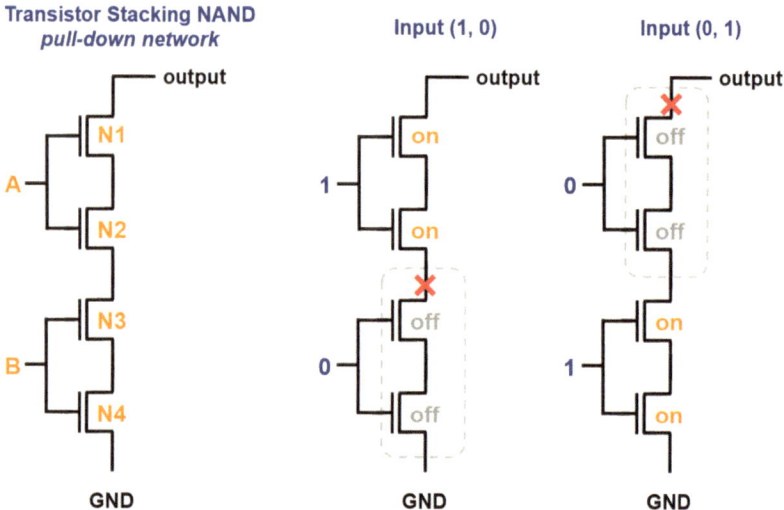

Fig. 5.14 Pull-down transistor network of the TS-based NAND gate and the implications of input (1, 0) and (0, 1). Particle hits are represented by a red cross

stacking structure of each gate. In Fig. 5.14, the pull-down network of the TS-based NAND is shown, containing the 4-stacked NMOS transistors. Considering the input (1, 0), which no SET was observed in the output, the sensitive transistors are placed next to the ground supply and far from the output. In this case, whenever a particle hits the *off* transistor (red cross), the SET pulse is electrically masked by the 2-stacked transistor series before reaching the output of the gate. On the other hand, for the input (0, 1), the *off* transistors are placed just next to the output of the gate. Thus, whenever a particle deposits sufficient charge near the *off* transistor next to the output, one SET will be observed. This same analogy can be drawn to the NOR gate where the 4-stacked PMOS transistors will mask any SET from the transistors placed next to the VDD supply. In summary, when adopting transistor stacking: (1) the transistors placed far from the output of the gate will be very likely hardened to any SET due to the electrical masking effect inherent of the stacking structure; (2) the worst-case input scenario is worsened due to the reduced driving capability of the series transistors in the stacking structure.

5.3 Part II—Transistor Folding (TF) and Diffusion Splitting (DS)

The transistor folding layout technique is a common practice in both digital and analog circuit designs, aimed at enhancing performance and regularity in VLSI circuits [15]. For instance, when larger transistors are required, exceeding the fixed

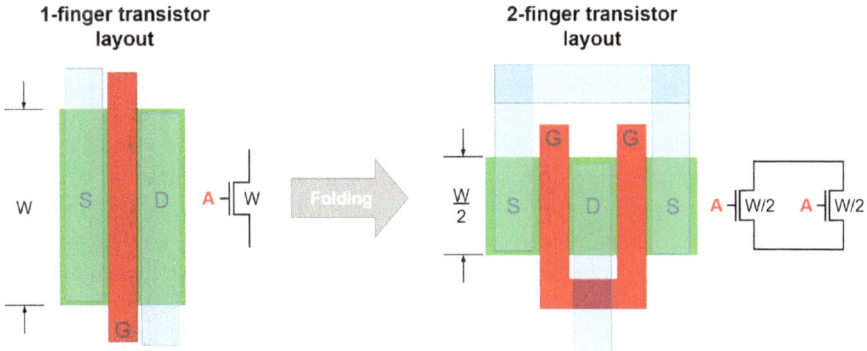

Fig. 5.15 Transistor folding layout technique

cell height of a given circuit design, the folding technique is employed. This method involves connecting parallel transistors with reduced channel width to achieve a larger overall width. For a transistor with a channel width W, the same W/L ratio, and thus the same drive strength, can be achieved by connecting n transistors with channel widths equal to W/n. Figure 5.15 illustrates the principle of folded transistor layouts, particularly in the case of a double-finger transistor. This layout technique divides the drain and source area into smaller partitions, potentially resulting in a significant reduction in drain area. For example, instead of using the SDS connection shown in Fig. 5.15, the circuit designer could opt for the DSD connection and limit the reduction in drain area.

In terms of radiation effects, this technique offers a reduced collecting drain area while preserving the same drive strength. The transistor size of the folded designs can be computed using Eq. 5.2, where W_F represents the width of each folded transistor, N_F denotes the number of fingers, and W stands for the transistor width in the original design.

$$W_F = \frac{1}{N_F} \times W. \tag{5.2}$$

The transistor folding technique can be combined with other hardening techniques such as gate sizing or dummy transistors/gates. The work in [16] was the first to propose transistor folding along with gate sizing to harden a circuit against both SEUs and SETs. A 3D mixed-mode TCAD (Technology Computer-Aided Design) simulation was carried out to analyze the SET pulse characteristics considering alpha particle and heavy ions hit on the center of the drain junctions [16]. Different from the sizing approach, which increases the circuit drive strength at the cost of increased drain area, transistor folding is able to reduce the transistor drain area while keeping the same drive strength. The transistor sizing was able to improve the robustness only for low-energy particles, while the transistor folding showed also improvement when considering high-energy particles [16, 17]. In [18], different well structures and layout topologies were studied to evaluate the Propagation-Induced Pulse Broadening (PIPB) effect in inverter chains. Accordingly, a double-finger inverter chain was compared against a single-finger inverter chain. In agreement with [16], the heavy-ion results show a reduction on the overall SET pulse width, but minimum influence in the PIPB effect. Inverter chains hardened with guard rings were also evaluated using single- and double-finger layout configurations with laser irradiation in [19]. Again, results showed an insignificant pulse broadening factor for the folded inverter chain; however, a wider SET pulse width average was observed in this case [19]. The authors attributed this to the larger spacing between the drain junctions and the guard rings in the folded design, which limits the charge collection reduction provided by the guard rings. Most studies have focused on inverter chain analysis, primarily with 2-finger layout configurations. In this section, transistor folding is applied to inverter, NAND, and NOR gates, with analysis conducted through layout-based predictive Monte Carlo simulations [17, 20].

The target circuit layouts were fully designed following a commercial Process Design Kit (PDK) in a bulk 65 nm technology. Additionally, all circuits are compatible with a standard cell library approach. Minimum width, spacing, and alignment/symmetry of each layer are carefully addressed to provide compatibility among the standard cells of the target technology. The cell height is set to 13 tracks of metal pitch, i.e., 2.6 μm high. To provide flexibility in cell routing, the metal 1 (M1) is primarily used for the intracell connections, except for some cases in which metal 2 (M2) is used horizontally. The PMOS transistor width is 760 nm, while the NMOS transistor width is 540 nm. To analyze exclusively the impact of the folded layouts, the equivalent gate sizing was kept the same for all cases. After all circuit designs are DRC (Design Rule Check) clean, LVS (Layout Versus Schematic) checked, and logic and electrical characterization is performed, the collecting drain area information can be extracted from the GDS (Graphical Design System) format file and submitted to the MC-Oracle tool [21]. All circuits were driving a fan-out 1 (FO1), i.e., an inverter was coupled to its output signal. Only the SET pulses with peak voltage higher than 0.6 V (half the supply voltage) are considered for the cross-section calculation. In addition to the double-finger designs (Fig. 5.15), quadruple-finger layout configurations are also considered in this work as shown in Fig. 5.16. However, one of the drawbacks of increasing the number of fingers in the folding technique, while maintaining the same gate sizing, is the increase in layout area due to the misuse of the fixed cell height. Thus, to address the area overhead associated with multiple-finger designs, Diffusion Splitting (DS) layout technique was first proposed in [17] and proven to enhance the radiation robustness of the digital circuits. As shown in Fig. 5.16, instead of using a single strip of active diffusion, a 2-row stacked diffusion transistor can be used. In [22], TCAD simulations demonstrated the efficacy of DS-based layouts, termed Splitting Active Area (SAA) layouts by the authors, in reducing charge collection efficiency, particularly for PMOS devices. To deepen this analysis, the impact on the in-orbit SET rates for heavy ions and protons is also presented later in this chapter.

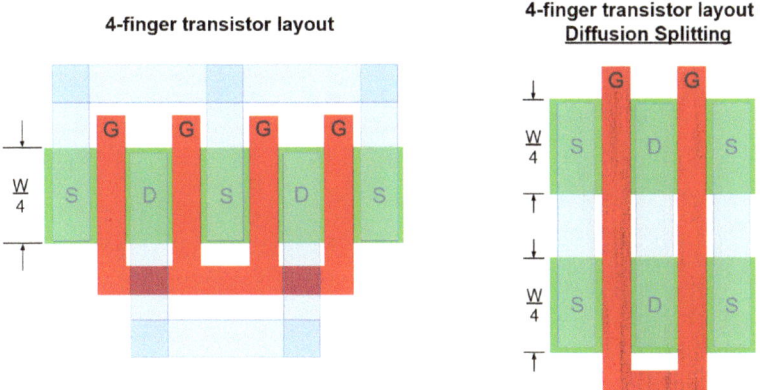

Fig. 5.16 Standard quadruple-finger folded transistor layout vs. folded transistor layout with Diffusion Splitting (DS) technique in which the diffusion strip is split into two strips and placed vertically aligned within each other

Notes
Besides reducing the area overhead of the folding technique, Diffusion Splitting (DS) technique improves the metal connection routability maintaining the same W/L ratio and the number of gate fingers.

5.3.1 Impact on the SET Cross Section

For the three cells analyzed, the double-finger layout configuration exhibited an area increase of approximately 1.5× the original unhardened single-finger layout area. An area increase around 2.5× is expected when using quadruple-finger layout designs. However, if DS is used, the area overhead for the four-finger designs can be reduced to the same observed in the two-finger designs, that is, approximately 1.5× greater than the original designs. Thus, DS provides an area reduction of 36% and 42% for the four-finger inverter and NAND/NOR cells, respectively. Once again, both NAND and NOR gates provide the same layout area in the original and folded designs due to the layout design regularity inherent of standard cell libraries. In Fig. 5.17, the simplified layouts of the NAND designs containing only M1, diffusion, and poly layers are shown. In standard cell methodology, a fixed cell height is defined to provide regularity. Thus, as the number of fingers is increased, the cell width is increased and, consequently, the cell layout increases. As the transistor sizing is kept the same, the DS can be applied to reduce the impact on the layout area, as shown in Fig. 5.17. For the sake of compactness, the layout designs for the inverter and NOR gates are omitted.

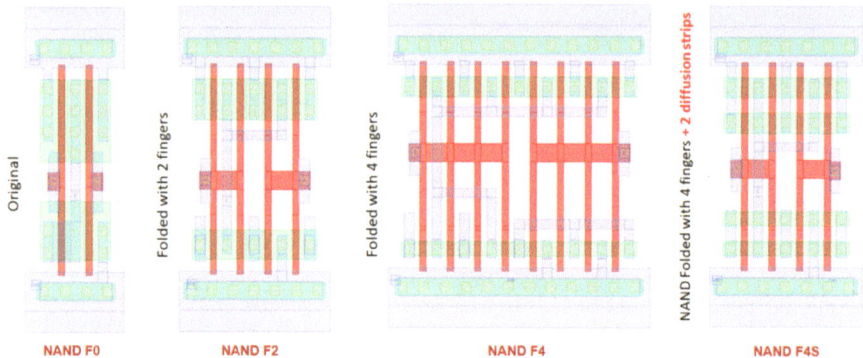

Fig. 5.17 Simplified layout design of NAND F0 (no folding, unhardened), NAND F2 (2-finger design), NAND F4 (4-finger design), and NAND F4S (4-finger design with diffusion splitting). For clarity, only the metal1 (blue), active diffusion (green), and poly layers (red) are shown

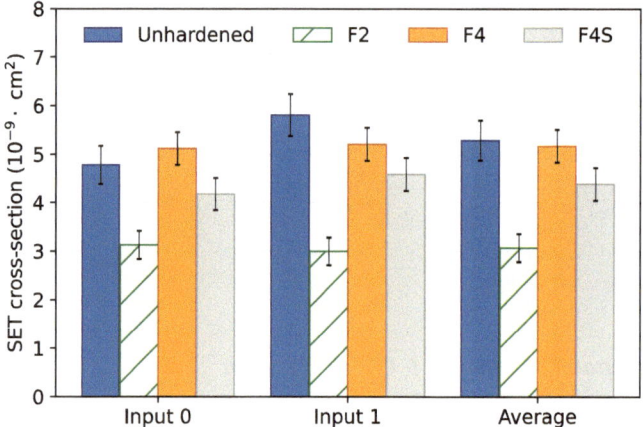

Fig. 5.18 SET cross section for the inverter designs under LET $= 78.23$ MeV.cm^2/mg considering input 0, input 1, and the mean value

The SET cross section σ_{SET} for the inverter designs under particle LET of 78.23 MeV.cm^2/mg is shown in Fig. 5.18. The folded designs have shown similar σ_{SET} for input 1 and input 0. Thus, using transistor folding may reduce the SET sensitivity dependence on the input signal in the inverter design at high LET. On average, the folded designs provide lower sensitivity than the unhardened design with the greatest σ_{SET} reduction for the 2-finger layout configuration, approximately 42%. The 4-finger inverter shows improvement solely for the input 1; however, despite the area reduction, DS also reduced the σ_{SET}. To analyze the folding impact for low particle LET irradiation, Fig. 5.19 presents the SET cross section for the inverter designs considering heavy ions with LET $= 5.43$ MeV.cm^2/mg. In this case, the greatest σ_{SET} reduction is observed for the 4-finger inverter with DS, about 37%.

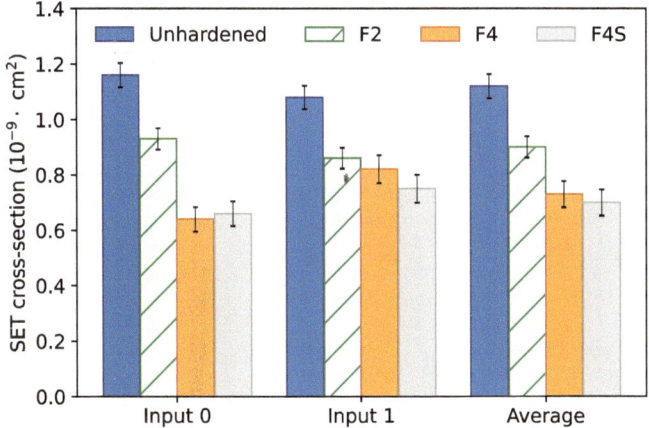

Fig. 5.19 SET cross section for the inverter designs under LET $= 5.43$ MeV.cm^2/mg considering input 0, input 1, and the mean value

It can be observed that, for low LET, the σ_{SET} reduces with the increase of the number of fingers N_F, in agreement with 3D TCAD results obtained in [16]. As N_F increases, the collecting regions are reduced and sparsely distributed along the layout, and then the charge sharing at low LET is limited. Thus, less folded transistors are affected by a single particle hit, leading to an improvement in the efficiency of the technique in reducing the σ_{SET}.

Considering the NAND designs, Fig. 5.20 presents the SET cross section σ_{SET} for particle LET equals to 78.23 MeV.cm^2/mg. Except for the 2-finger design, the folded designs provided a higher mean σ_{SET}. At high LET, the hardening efficiency of folded transistors is limited due to the complex input dependence observed on the cross section. The folded NAND designs show a stronger input dependence than the original unhardened version, leading to a similar or lower σ_{SET} only for inputs (0, 0) and (0, 1). The worst-case input scenario for NAND gates is the input (1, 1), and, in this case, transistor folding exacerbates the SET sensitivity up to approximately 62% in the 4-finger design.

As observed for the gate sizing and transistor stacking designs in the previous section, the layout-based hardening techniques can worsen the SET robustness in the worst-case input scenario. For input (1, 1), all PMOS devices are sensitive to a particle hit, and besides the lower restoring capability of NMOS devices, PMOS transistors collect more charge due to its larger drain area [23]. Similarly, in the NOR case, the input (0, 0) is the worst-case input scenario, and it shows increased σ_{SET} when adopting folded transistors as shown in Fig. 5.21. To reduce the increased sensitivity at the worst-case input scenarios, transistor folding can be applied only in the pull-down (or pull-up) network to balance the overall SET sensitivity. The asymmetric designs are also considered in this chapter, and it is discussed in the following section. Despite the poor hardening performance for high particle LET, the folded designs have shown great reduction on the overall SET cross section for

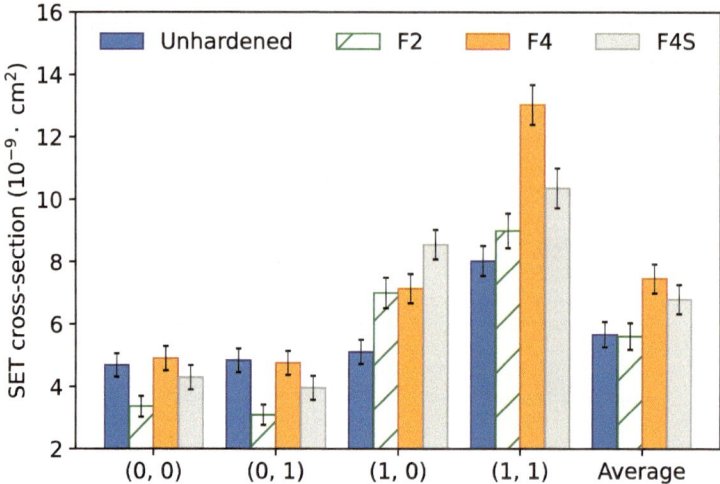

Fig. 5.20 SET cross section for the NAND designs at LET $= 78.23$ MeV.cm^2/mg for each input signal and the mean value

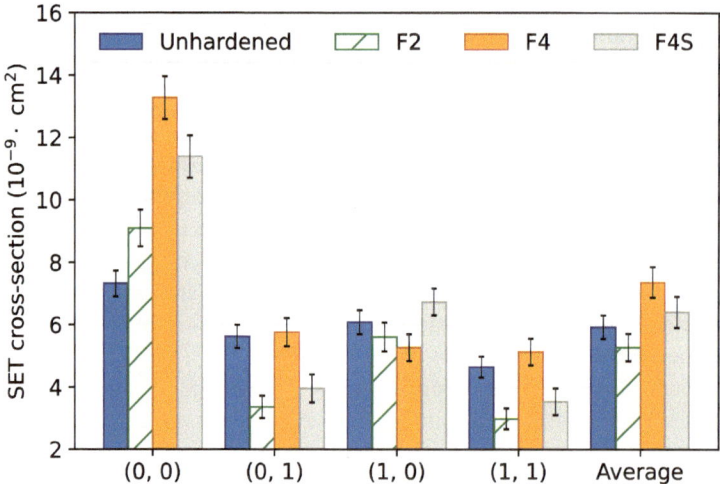

Fig. 5.21 SET cross section for the NOR designs at LET $= 78.23$ MeV.cm^2/mg for each input signal and the mean value

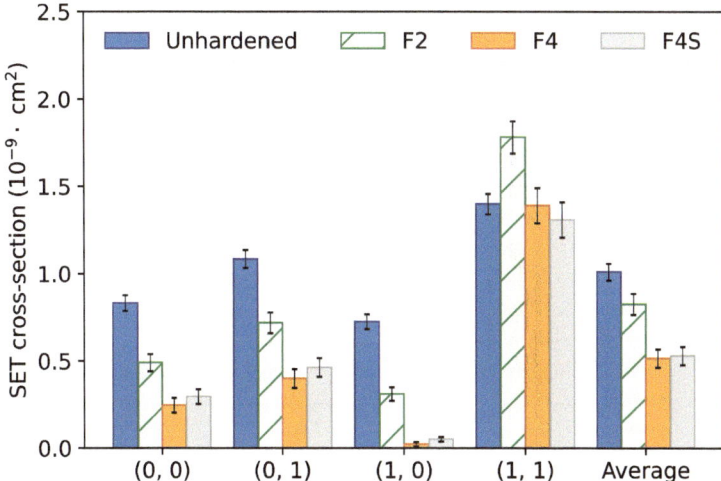

Fig. 5.22 SET cross section for the NAND designs at LET $= 5.43$ MeV.cm^2/mg for each input signal and the mean value

low LET. Considering the NAND design under particle LET of 5.43 MeV.cm^2/mg shown in Fig. 5.22, the 4-finger design showed the lowest mean σ_{SET}, and similar sensitivity is obtained with DS. Thus, it is important to note that, for low LET, increasing the N_F also improves the overall SET cross section σ_{SET} as observed for the inverter design.

However, the worst-case input scenario is still worsened by the technique, for instance, in the F2 NAND design. The full understanding of the impact on the SET cross section can be better visualized through the curves in Fig. 5.23. As previously mentioned, for high LET, the folded designs may exhibit similar or even worse radiation robustness than the original unfolded version (F0 designs). However, as the LET decreases below 10 MeV.cm^2/mg, the folded designs start to demonstrate improved or comparable cross sections to the observed for the F0 designs. For LET lower than 10 MeV.cm^2/mg, the 4-finger design (F4) is preferred. Except for the 2-finger NAND (F2), all other folded designs provided a higher threshold LET than the unfolded design.

5.3.2 Asymmetric Designs

As observed in the last subsection, transistor folding can induce a higher SET cross section for the worst-case input scenario of the NAND and NOR circuits. Thus, in this section, a deep analysis is provided in order to enable a better usage of the TF technique. Figure 5.24 presents the transistor network of a NAND gate along with its corresponding truth table. For the worst-case scenario, highlighted in red, the *off* transistors, which are sensitive to particle hits, are issued from the pull-up network, i.e., PMOS devices.

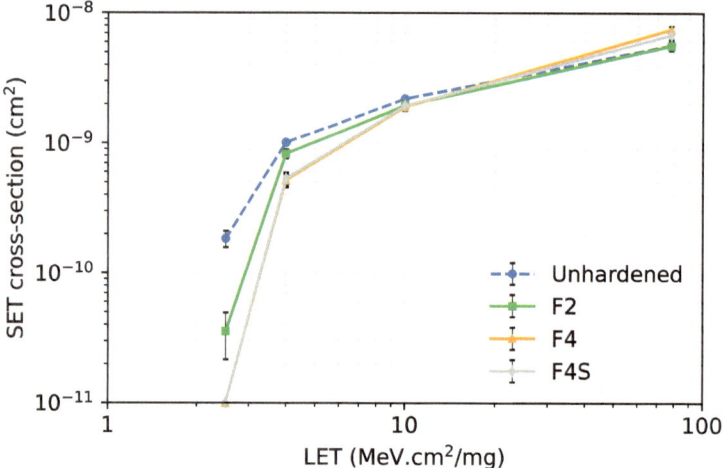

Fig. 5.23 Log–log representation of the mean SET cross-section curves for the NAND designs (F0: unfolded, F2: 2-finger, F4: 4-finger, F4S: 4-finger design with diffusion splitting). No event was observed in the F4 and F4S designs (triangle and rhombus curves) for LET lower than 5.43 MeV.cm^2/mg

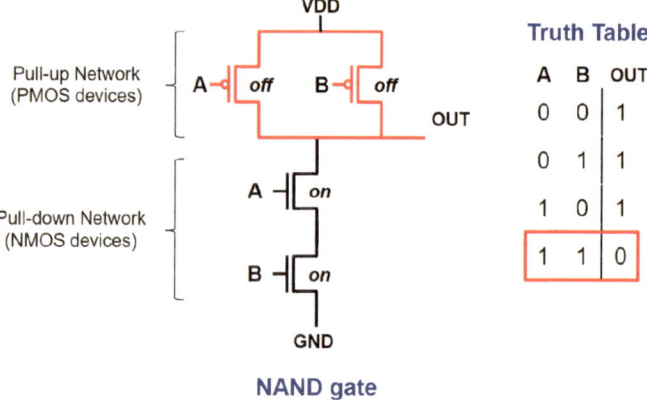

Fig. 5.24 Transistor network of an NAND gate and its truth table. *On* state and *off* state are indicated for the transistors considering the worst-case input scenario, i.e., input (1, 1)

So far, the folding technique has been equally applied to both pull-up and pull-down networks. However, in order to investigate the impact of asymmetric designs, this analysis explores two specific configurations: the 2-finger design and the 4-finger design with drain/source (DS) adopting folding in only one of the networks, while the other network remains unfolded. By examining the effects of asymmetric designs, a more comprehensive understanding of the impact of transistor folding on SET cross sections can be obtained. This analysis aims to determine the effectiveness of selectively applying the folding technique to either

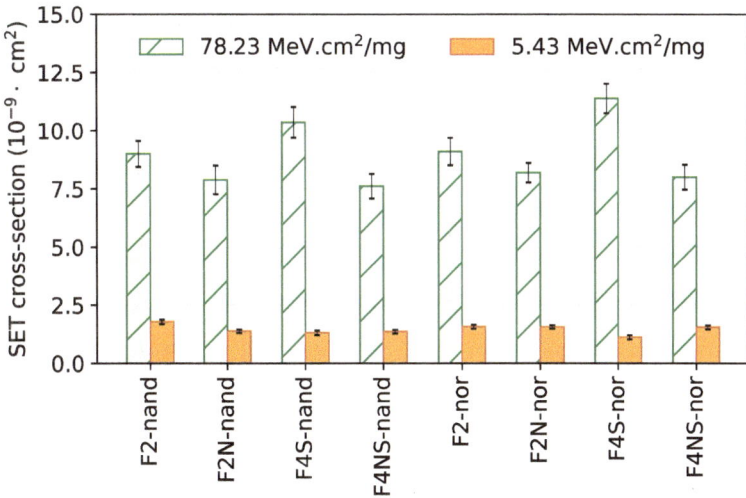

Fig. 5.25 SET cross section for symmetric and asymmetric designs of 2-finger and 4-finger with diffusion splitting of NAND and NOR gates in the worst-case input scenario, i.e., input (1, 1) and input (0, 0), respectively

the pull-up or pull-down network, taking into account the specific characteristics of the worst-case scenario.

The folding technique was applied only in the NMOS devices for the NAND gate, and only in the PMOS devices for the NOR gate. The SET cross sections are shown in Fig. 5.25 for each design considering only the worst-case input scenario, i.e., the input combination (1, 1) and (0, 0) for the NAND and NOR, respectively. The number indicates the fingers, N/P indicates when only NMOS/PMOS devices are folded, and S indicates when diffusion splitting is adopted. For instance, the F4NS circuit is the 4-finger design with only NMOS devices folded and with diffusion splitting. This nomenclature is used in the remaining of this chapter for the sake of compactness. For the LET $= 78.23$ MeV.cm^2/mg, the asymmetric designs were able to improve the SET robustness of the circuits. The greatest reduction on the cross section was observed for the 4-finger designs. The NAND F4NS circuit provides a reduction of 26.5% when compared to the NAND F4S circuit, while the NOR F4PS circuit has approximately 29.7% of reduction compared to the NOR F4S.

However, when adopting the asymmetric designs, not only the worst-case input SET cross section is affected but in all input scenarios. It can be seen in Figs. 5.26 and 5.27 in which the SET cross section σ_{SET} for each input signal and the mean value are shown for the NOR and NAND gate, respectively.

After examining the results of the asymmetric designs for NOR and NAND circuits, it is evident that there is a slight increase in the cross section for specific input combinations, such as (0, 1), (1, 0), and (1, 1). For the NOR F4PS circuit, the mean SET cross-section reduction is approximately 13%, while the NOR F2P circuit demonstrates a similar mean SET cross section to the NOR F2 circuit. Similar observations can be made for the NAND designs, as depicted in Fig. 5.27.

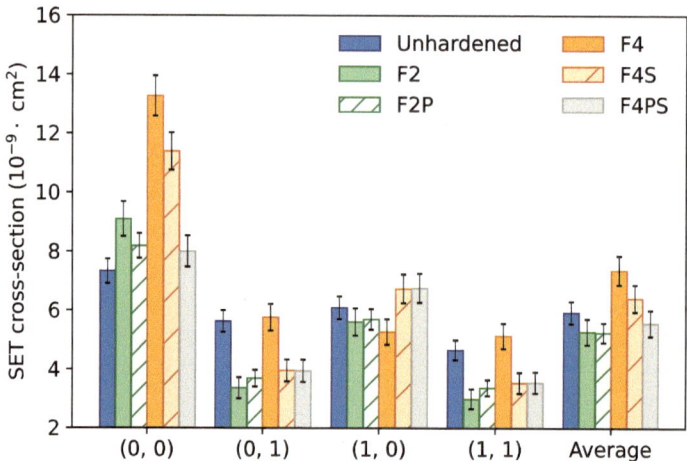

Fig. 5.26 SET cross section for the NOR designs at LET $= 78.23$ MeV.cm^2/mg for each input signal and the mean value

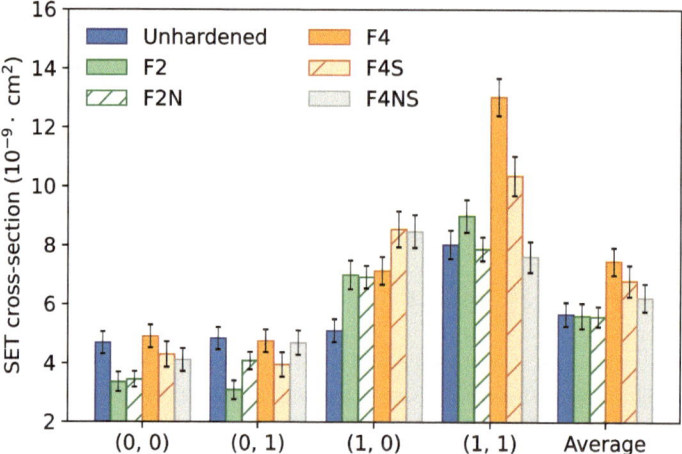

Fig. 5.27 SET cross section for the NAND designs at LET $= 78.23$ MeV.cm^2/mg for each input signal and the mean value

These findings reinforce the importance of considering input dependencies when implementing radiation hardening techniques in digital logic circuits.

5.3.3 Voltage Variability

In addition to radiation effects, electronic circuits are also vulnerable to dynamic environmental variability, including voltage fluctuations caused by voltage drops and di/dt noise in current derivatives [24]. Previous studies have highlighted that

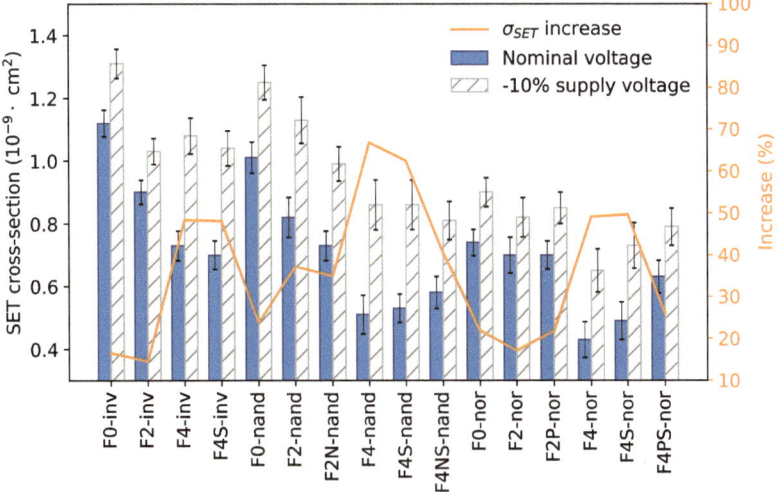

Fig. 5.28 SET cross section considering a voltage drop of 10% over the nominal supply voltage and a particle LET of 5 MeV.cm^2/mg. The percentage increase $\Delta\sigma_{SET}$ in the cross section is also shown in orange line

voltage variability can significantly impact circuit reliability, particularly in harsh environments where radiation effects are present [25, 26]. For instance, in [25], it was demonstrated that voltage variability reduces the threshold LET for various XOR topologies in FinFET and Trigate devices. Moreover, [26] observed a decrease in the electrical masking capability of gates and an increase in the SET pulse width under voltage variability conditions. Therefore, to comprehensively assess the impact of transistor folding, diffusion splitting, and asymmetric designs, the influence of voltage variability on SET cross section is evaluated and shown in Fig. 5.28.

The IR drops due to the parasitic resistance of the power grids can lead to ±10% variation on the supply voltage. Accordingly, all circuits were analyzed considering a voltage drop of −10% of the nominal supply voltage of the technology, i.e., 0.12 V. The SET cross section and the variation $\Delta\sigma_{SET}$ (in %) are shown for each circuit in Fig. 5.28. As expected, a reduction on the supply voltage of the circuits reduces the driving capability and consequently reduces the recovery efficiency, leading to a higher σ_{SET}. The usage of diffusion splitting technique induces insignificant impact on the circuit robustness to voltage variability as similar SET cross sections are obtained for the F4 and F4S circuits. However, for the 3 logic gates (inverter, NAND and NOR), the F4 and F4S circuits showed the greatest σ_{SET} increase, ranging from 50% to 70%. Thus, as the number of fingers is increased, the circuit becomes more sensitive to the voltage drops. However, the asymmetric designs have shown to reduce the impact of voltage variability on the SET cross sections. Considering only the inverter designs, the F4S circuit shows the lowest σ_{SET} at nominal voltage, but it is the most sensitive to voltage drops, along with the F4 circuit. On the other hand, the F2 circuit provides the lowest σ_{SET} at -10% of the supply voltage, and also the lowest variation on σ_{SET}.

For the NAND gate, the F4 circuit provides the lowest σ_{SET} at nominal voltage, but it increases up to 66% with 10% reduction on the supply voltage. In this case, the unhardened circuit is the least sensitive design to voltage variation, leading to approximately 24% of increase in the σ_{SET}. However, under voltage drop, the lowest σ_{SET} is observed for the asymmetric design F4NS circuit. Compared to the unhardened NAND circuit, when the voltage fluctuation is considered, the F4NS reduced the σ_{SET} up to 35%. For the NOR gate, similarly to the NAND gate, the F4 circuit provides the lowest σ_{SET} at nominal voltage, but high sensitivity to the voltage drops, resulting in approximately 49% increase in the σ_{SET}. However, it still provides the lowest σ_{SET} during voltage fluctuation. The diffusion splitting used in F4S increased its cross section, but it still provides a lower σ_{SET} than the F2 circuit, with the same area overhead. This initial analysis indicates that variability should be carefully taken into consideration when adopting layout techniques as it is critical for circuit-based techniques.

Process, Voltage, and Temperature variability, also known as **PVT variability**, will also affect the sensitivity of your design and should be considered in the design phase.

5.3.4 Impact on the In-Orbit SET Rate: LEO and ISS Orbits

To study the impact of adopting these techniques in a radiation environment, the in-orbit SET rates were estimated for the Low-Earth Orbit (LEO) and International Space Station (ISS) orbits. The OMERE software was used based on the SET cross-section curves calculated with the current database provided by MC-Oracle. OMERE is a tool dedicated to the analysis of space environment and radiation effects on electronics developed by TRAD and CNES [27]. The Integral Rectangular Parallelepiped (IRPP) approach is used to predict the SET rate, i.e., it is calculated by the convolution of the heavy-ion cross-section data with the particle flux in the aforementioned orbits. This approach is a standard method specified by the European Cooperation for Space Standardization (ECSS) [28]. The NASA AP8MIN trapped radiation model is used for the proton fluxes under solar minimum conditions [29]. For the Galactic Comic Rays (GCR) fluxes, the international standard ISO 15390 model is used [30]. A fixed shielding of 1 g/cm^2 is considered. The calculated heavy-ions SET rates are shown in Fig. 5.29. Firstly, it can be noticed that all folded designs exhibit lower rates than its unhardened version (F0 circuit) for both orbits. In the case of the inverter, the F4S circuit provides the lowest rate with a reduction of approximately 82% and 77% in the ISS and LEO orbits, respectively. Similarly, for the NOR gate, the F4S showed the lowest SET rate. The

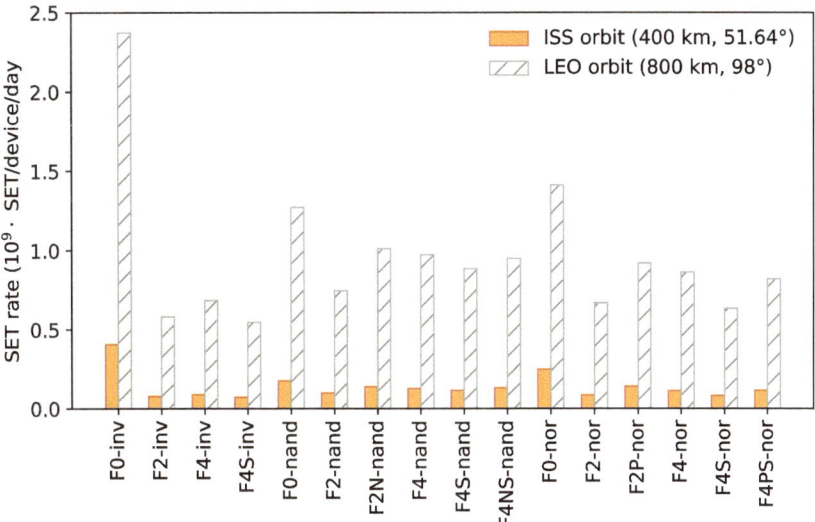

Fig. 5.29 Heavy-ions SET rate estimated with OMERE [27] for each circuit and its hardened version considering the LEO (800 km, 98°) and ISS (400 km, 51.64°) orbit. The SET cross-section average of each input signal is used

folded transistor and diffusion splitting provided about 66% and 55% of reduction on the SET rate for the ISS and LEO orbits, respectively.

On the other hand, the F2 circuit is expected to have a lower rate for the NAND gate. Although the cross-section calculations were performed for heavy ions, the protons are expected to dominate the SEE rates in the LEO and ISS orbits. Due to its improved accuracy when compared to analytical models, the METIS method [31–33] was used to predict the proton-induced SET cross-section curves from the heavy-ions data. The sums of the SET rate induced by heavy ions and protons are shown in Fig. 5.30. As expected, the overall SET rate increased considerably. However, in this case, the F4 and F4S circuits are no longer the most hardened designs. The F2 circuits have shown the lowest rate for the inverter and NAND gate, for both orbits. For the NOR gate, the asymmetric design F2P provided the lowest rate, about 10% reduction for the two orbits. Except for the F2N, one can notice that whenever the asymmetric design approach is adopted, a reduction on the overall SET rate is observed.

5.3.5   Transistor Scaling and Angular Dependence

As the demand for improved performance and low power in critical systems grows, future missions are considering advanced technology nodes. However, the effectiveness of hardening techniques may diminish with transistor scaling, as close proximity increases charge sharing effects. Nevertheless, in FinFET technologies,

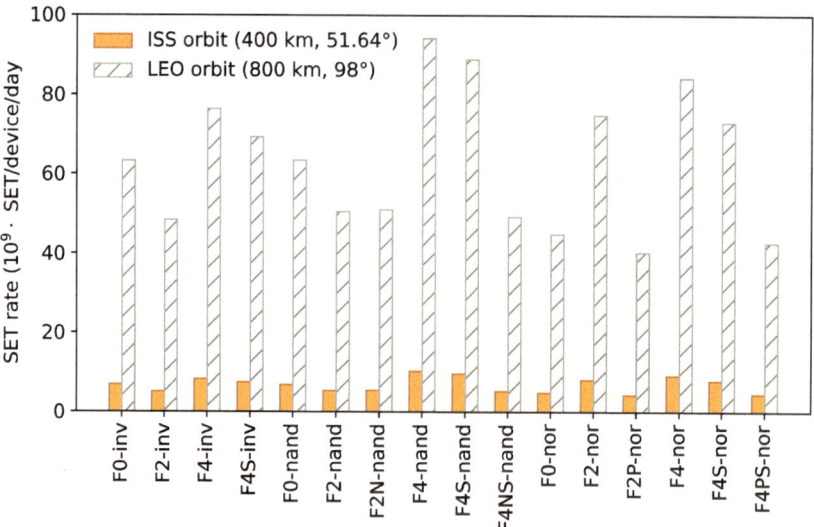

Fig. 5.30 Total SET rate estimated with OMERE [27] for each circuit and its hardened version for the LEO (800 km, 98°) and ISS (400 km, 51.64°) orbits considering heavy ions and protons. Proton-induced SET rates are estimated with the METIS method [31]. The SET cross-section average of each input signal is used

the three-dimensional structure of the transistor impacts on the charge collection process, and a reduction on the charge sharing effects might be expected as shown in [34]. Additionally, adoption of SOI technology could significantly enhance hardening efficiency by eliminating diffusion contributions to charge collection [35].

Regarding angular effects on charge collection, research indicates that folded transistors exhibit minimal angle dependence [36]. It was shown that only the NMOS transistors experienced an increase in collected charge for angled strikes. For the worst-case scenario, in which the particle strikes the center of the drain area with a tilt angle of 60°, the total collected charge of a 4-finger NMOS transistor can be 20% higher than the collected charge in the original design, without transistor folding. Thus, a very weak angle dependence can be expected in the SET cross section of the folded designs, especially for low LET due to the reduced charge sharing between adjacent devices. Further, according to [37], deeply scaled CMOS technologies present a marginal difference on the overall charge sharing effect between normal and angled strikes.

5.4 Summary

Physical layout design influences the main SEE mechanisms in VLSI circuits such as charge collection and charge sharing effects. Accordingly, RHBD techniques can be adopted at layout level to improve the radiation robustness of electronic circuits. In this chapter, three RHBD techniques exploring layout modifications were analyzed under heavy ions: gate sizing, transistor stacking, and transistor folding. Firstly, the gate sizing and transistor stacking were studied based on pre-designed standard cells. The idea is to investigate how conventional standard cell libraries can be used to maximize the reliability of VLSI systems under radiation effects. Besides the area and leakage current increase, both techniques were able to provide a reduction on the overall SET cross section, especially for low particle LET. The NOR gate shows the greatest improvements on the SET cross section even though transistor stacking can increase the maximum SET pulse width to $2\times$ wider than the original design. Gate sizing shows the best trade-off between area, power, and reliability. However, the hardening efficiency of transistor stacking is strongly dependent on the input signal of the gate. This is a reflection to the fact that, in the stacking structure, the transistors placed far from the output of the gate will be likely unable to induce a SET pulse in the output due to the electrical masking effect. Thus, according to the application, this technique can possibly outperform gate sizing. In the next chapter, signal probability will be used in order to enhance the hardening efficiency of RHBD techniques.

After understanding the implications of adopting gate sizing and transistor stacking, the efficiency of transistor folding layout in improving the SET immunity of digital circuits was presented. The results have shown that folded designs can provide lower SET cross section in addition to the higher threshold LET than the observed for the unfolded designs. The number of fingers was also explored. At high LET, the 2-finger designs showed the best performances. However, for LET lower than $10\,\mathrm{MeV.cm^2/mg}$, the hardening efficiency of the folded designs is expected to increase as the number of fingers is increased. Increasing the number of fingers increases greatly the final layout area. Thus, a layout technique was proposed to overcome the area overhead of multiple-finger designs. In the diffusion splitting approach, the active diffusion is split into two strips and placed vertically aligned within each other. Besides reducing the layout area, diffusion splitting may also improve the SET cross section depending on the circuit topology, input signal, and ion LET. Due to the strong input dependence of these techniques, it was also proposed to adopt asymmetric designs, i.e., applying the hardening techniques only in the PMOS or NMOS devices, depending on the worst-case input scenario of the logic gate. Voltage variability was also explored due to its impact on the reliability of deeply scaled technologies. The folded designs have shown a higher sensitivity to voltage fluctuation. However, the usage of asymmetric designs showed to reduce it. And, lastly, the in-orbit SET rates were predicted for the LEO and ISS orbits. When the SET rate is only calculated for heavy ions, all the folded designs provided lower rate than the unfolded designs. However, the protons dominate the SEE rates in the LEO and ISS orbits. When protons are taken into account, the 2-finger designs (symmetric and asymmetric) and the asymmetric 4-finger design with DS are the most hardened circuits. Overall, the chapter provides insights into the impact of layout-level RHBD techniques on radiation robustness and explores the potential trade-offs and considerations for their practical implementation.

Highlights

- Physical layout design has a significant impact on the fundamental Single-Event Effects (SEE) mechanisms in VLSI circuits, thereby providing an opportunity to enhance radiation robustness through design-based hardening techniques at the layout level.
- Gate sizing and transistor stacking were analyzed as RHBD techniques, with both techniques reducing the overall SET cross section and improving reliability, although the efficiency of transistor stacking is highly dependent on the input signal.
- The evaluation of transistor folding layouts revealed their potential in enhancing SET immunity by achieving lower SET cross sections and higher threshold Linear Energy Transfer (LET) compared to unfolded designs. The number of fingers utilized in the layout design plays a significant role in the hardening efficiency, with 2-finger designs performing well at high LET values.
- Diffusion splitting was proposed as a technique to reduce layout area in multiple-finger designs, with potential SET cross-section improvements depending on circuit topology, input signal, and ion LET.
- Asymmetric designs, applying hardening techniques to specific types of devices, can enhance the efficiency of certain hardening techniques.
- When evaluating the efficiency of hardening techniques, it is essential to consider variability factors, such as voltage fluctuations and process variations, to ensure robustness in real-world scenarios.

References

1. Daniel B Limbrick, Nihaar N Mahatme, William H Robinson, and Bharat L Bhuva. Reliability-aware synthesis of combinational logic with minimal performance penalty. *IEEE Transactions on nuclear science*, 60 (4): 2776–2781, 2013.
2. Bradley T Kiddie and William H Robinson. Alternative standard cell placement strategies for single-event multiple-transient mitigation. In *2014 IEEE Computer Society Annual Symposium on VLSI*, pages 589–594. IEEE, 2014.
3. Yankang Du, Shuming Chen, and Biwei Liu. A constrained layout placement approach to enhance pulse quenching effect in large combinational circuits. *IEEE Transactions on Device and Materials Reliability*, 14 (1): 268–274, 2013.
4. L. Entrena, A. Lindoso, E. S. Millan, S. Pagliarini, F. Almeida, and F. Kastensmidt. Constrained placement methodology for reducing SER under single-event-induced charge sharing effects. *IEEE Transactions on Nuclear Science*, 59 (4): 811–817, Aug 2012. ISSN 0018-9499. doi: https://doi.org/10.1109/TNS.2012.2191796.
5. Olivier Coudert. Gate sizing for constrained delay/power/area optimization. *IEEE Transactions on Very Large Scale Integration (VLSI) Systems*, 5 (4): 465–472, 1997.

6. Gracieli Posser, Guilherme Flach, Gustavo Wilke, and Ricardo Reis. Gate sizing minimizing delay and area. In *2011 IEEE Computer Society Annual Symposium on VLSI*, pages 315–316. IEEE, 2011.
7. Quming Zhou and Kartik Mohanram. Gate sizing to radiation harden combinational logic. *IEEE Transactions on Computer-Aided Design of Integrated Circuits and Systems*, 25 (1): 155–166, 2005.
8. James E Stine, Ivan Castellanos, Michael Wood, Jeff Henson, Fred Love, W Rhett Davis, Paul D Franzon, Michael Bucher, Sunil Basavarajaiah, Julie Oh, et al. FreePDK: An open-source variation-aware design kit. In *2007 IEEE international conference on Microelectronic Systems Education (MSE'07)*, pages 173–174. IEEE, 2007.
9. Ethan H Cannon and Manuel Cabanas-Holmen. Heavy ion and high energy proton-induced single event transients in 90 nm inverter, NAND and nor gates. *IEEE Transactions on Nuclear Science*, 56 (6): 3511–3518, 2009.
10. Ygor Q. Aguiar, Laurent Artola, Guillaume Hubert, Cristina Meinhardt, Fernanda Kastensmidt, and Ricardo Reis. Evaluation of radiation-induced soft error in majority voters designed in 7 nm FinFET technology. *Microelectronics Reliability*, 2017a. doi:10.1016/j.microrel.2017.06.077.
11. JS Kauppila, TD Loveless, RC Quinn, JA Maharrey, ML Alles, MW McCurdy, RA Reed, BL Bhuva, LW Massengill, and K Lilja. Utilizing device stacking for area efficient hardened SOI flip-flop designs. In *2014 IEEE International Reliability Physics Symposium*, pages SE–4. IEEE, 2014.
12. H-B Wang, L Chen, R Liu, Y-Q Li, JS Kauppila, BL Bhuva, K Lilja, S-J Wen, R Wong, R Fung, et al. An area efficient stacked latch design tolerant to SEU in 28 nm FDSOI technology. *IEEE Transactions on Nuclear Science*, 63 (6): 3003–3009, 2016.
13. Kodai Yamada, Haruki Maruoka, Jun Furuta, and Kazutoshi Kobayashi. Sensitivity to soft errors of NMOS and PMOS transistors evaluated by latches with stacking structures in a 65 nm FDSOI process. In *2018 IEEE International Reliability Physics Symposium (IRPS)*, pages P–SE. IEEE, 2018.
14. Yibin Ye, Shekhar Borkar, and Vivek De. A new technique for standby leakage reduction in high-performance circuits. In *1998 Symposium on VLSI Circuits. Digest of Technical Papers (Cat. No. 98CH36215)*, pages 40–41. IEEE, 1998.
15. TW Her and DF Wong. Cell area minimization by transistor folding. In *Proceedings of EURO-DAC 93 and EURO-VHDL 93-European Design Automation Conference*, pages 172–177. IEEE, 1993.
16. F Lima Kastensmidt, T Assis, I Ribeiro, G Wirth, L Brusamarello, and R Reis. Transistor sizing and folding techniques for radiation hardening. In *2009 European Conference on Radiation and Its Effects on Components and Systems*, pages 512–519. IEEE, 2009.
17. Y. Q. Aguiar, Frédéric Wrobel, J-L Autran, FL Kastensmidt, P Leroux, F Saigné, V Pouget, and AD Touboul. Exploiting transistor folding layout as RHBD technique against single-event transients. *IEEE Transactions on Nuclear Science*, 67 (7): 1581–1589, 2020a.
18. Yaqing Chi, Ruiqiang Song, Shuting Shi, Biwei Liu, Li Cai, Chunmei Hu, and Gang Guo. Characterization of single-event transient pulse broadening effect in 65 nm bulk inverter chains using heavy ion microbeam. *IEEE Transactions on Nuclear Science*, 64 (1): 119–124, 2016.
19. Wen Zhao, Chaohui He, Wei Chen, Rongmei Chen, Peitian Cong, Fengqi Zhang, Zujun Wang, Chen Shen, Lisang Zheng, Xiaoqiang Guo, et al. Single-event multiple transients in guard-ring hardened inverter chains of different layout designs. *Microelectronics Reliability*, 87: 151–157, 2018.
20. Ygor Q. Aguiar, Frédéric Wrobel, Jean-Luc Autran, Paul Leroux, Frédéric Saigné, Vincent Pouget, and Antoine D Touboul. Mitigation and predictive assessment of set immunity of digital logic circuits for space missions. *Aerospace*, 7 (2): 12, 2020b.
21. Frédéric Wrobel and Frédéric Saigné. MC-ORACLE: A tool for predicting soft error rate. *Computer Physics Communications*, 182 (2): 317–321, 2011.

22. Bin Liang, Deng Luo, Qian Sun, and Wangyong Chen. Layout based radiation hardening techniques against single-event transient. *Microelectronics Reliability*, 135: 114572, 2022.
23. Y. Q. Aguiar, Frédéric Wrobel, S. Guagliardo, J-L Autran, P. Leroux, F. Saigné, A. D. Touboul, and V. Pouget. Radiation hardening efficiency of gate sizing and transistor stacking based on standard cells. *Microelectronics Reliability*, 100: 113457, 2019.
24. Shekhar Borkar, Tanay Karnik, Siva Narendra, Jim Tschanz, Ali Keshavarzi, and Vivek De. Parameter variations and impact on circuits and microarchitecture. In *Proceedings of the 40th annual Design Automation Conference*, pages 338–342. ACM, 2003.
25. Ygor Q. Aguiar, Cristina Meinhardt, and Ricardo AL Reis. Radiation sensitivity of XOR topologies in multigate technologies under voltage variability. In *Circuits & Systems (LAS-CAS), 2017 IEEE 8th Latin American Symposium on*, pages 1–4. IEEE, 2017b. doi: https://doi.org/10.1109/LASCAS.2017.7948075.
26. Semiu A Olowogemo, William H Robinson, and Daniel B Limbrick. Effects of voltage and temperature variations on the electrical masking capability of sub-65 nm combinational logic circuits. In *2018 IEEE International Symposium on Defect and Fault Tolerance in VLSI and Nanotechnology Systems (DFT)*, pages 1–6. IEEE, 2018.
27. The OMERE 5.3 software by TRAD and CNES. URL http://www.trad.fr/en/space/omere-software.
28. ECSS Secretariat. Space engineering: Calculation of radiation and its effects and margin policy handbook - ECSS-E-HB-10-12A. 2010.
29. Donald M Sawyer and James I Vette. AP-8 trapped proton environment for solar maximum and solar minimum. Technical report, National Aeronautics and Space Administration, 1976.
30. Aircraft Technical Committee ISO/TC 20, Space systems space vehicles, Subcommittee SC 14, and operations. Iso-15390: 2004. space environment (natural and artificial)-galactic cosmic ray model, 2004.
31. C. Weulersse, F. Wrobel, F. Miller, T. Carrière, R. Gaillard, J. R. Vaillé, and N. Buard. A Monte-Carlo engineer tool for the prediction of SEU proton cross section from heavy ion data. In *European Conference on Radiation and Its Effects on Components and Systems*, pages 376–383, Sept 2011. doi: https://doi.org/10.1109/RADECS.2011.6131348.
32. Cecile Weulersse, Sebastien Morand, Florent Miller, Thierry Carriere, and Renaud Mangeret. Simulation of proton induced set in linear devices and assessment of sensitive thicknesses. In *2015 15th European Conference on Radiation and Its Effects on Components and Systems (RADECS)*, pages 1–4. IEEE, 2015.
33. Cécile Weulersse, Sébastien Morand, Florent Miller, Thierry Carrière, and Renaud Mangeret. Predictions of proton cross-section and sensitive thickness for analog single-event transients. *IEEE Transactions on Nuclear Science*, 63 (4): 2201–2207, 2016.
34. Yi-Pin Fang and Anthony S Oates. Neutron-induced charge collection simulation of bulk FinFET SRAMs compared with conventional planar SRAMs. *IEEE Transactions on Device and Materials Reliability*, 11 (4): 551–554, 2011.
35. F El-Mamouni, EX Zhang, DR Ball, B Sierawski, MP King, RD Schrimpf, RA Reed, ML Alles, DM Fleetwood, D Linten, et al. Heavy-ion-induced current transients in bulk and SOI FinFETs. *IEEE Transactions on Nuclear Science*, 59 (6): 2674–2681, 2012.
36. Nicholas M Atkinson, AF Witulski, WT Holman, BL Bhuva, and JD Black. *Single-event characterization of a 90-nm bulk CMOS digital cell library*. PhD thesis, Vanderbilt University Nashville, 2010.
37. J. R. Ahlbin, M. J. Gadlage, D. R. Ball, A. W. Witulski, B. L. Bhuva, R. A. Reed, G. Vizkelethy, and L. W. Massengill. The effect of layout topology on single-event transient pulse quenching in a 65 nm bulk CMOS process. *IEEE Transactions on Nuclear Science*, 57 (6): 3380–3385, Dec 2010. ISSN 0018-9499. doi: https://doi.org/10.1109/TNS.2010.2085449.

Chapter 6
Analysis of Circuit-Based RHBD Techniques

6.1 Reliability-Driven Synthesis

The synthesis process is an essential stage in very-large-scale integration (VLSI) system design using integrated circuits (ICs) as it determines critical performance characteristics of the application, such as timing, power consumption, and area efficiency. However, this process also significantly impacts the final circuit's reliability, especially when considering radiation effects such as soft errors. Consequently, various approaches can be integrated into the synthesis flow to enhance this aspect. This chapter analyzes and proposes mitigation strategies as a novel radiation hardening by design (RHBD) approach applicable during both logical and physical synthesis in VLSI systems, offering significant improvements in circuit robustness under radiation effects.

Electronic circuit development for space and aviation can employ various design methodologies, ranging from field programmable gate arrays (FPGAs) to full-custom or cell-based application-specific integrated circuits (ASICs). While FPGA-based designs offer fast prototyping, they often sacrifice area and performance compared to full-custom designs [1, 2]. On the other hand, ASICs strike the best balance between performance, power consumption, and circuit area. The standard cell methodology is a primary approach in ASIC design, wherein thousands of predesigned and characterized logic gates, known as "standard cell logic gates," are utilized for designing complex VLSI circuits. The synthesis of a Boolean function can lead to varying combinations of logic cells, affecting the number of transistors and layout design, which directly influences the radiation robustness of the circuit. Once vulnerable nodes are identified within a circuit, hardening techniques such as gate sizing or hardware redundancy can be employed to enhance overall reliability [3, 4].

There is a growing trend within the research community to incorporate radiation hardening techniques early in the VLSI circuit design flow [5–10]. The proposed predictive single-event transient (SET) characterization methodology (discussed in

© The Author(s) 2025
Y. Quadros de Aguiar et al., *Single-Event Effects, from Space to Accelerator Environments*, https://doi.org/10.1007/978-3-031-71723-9_6

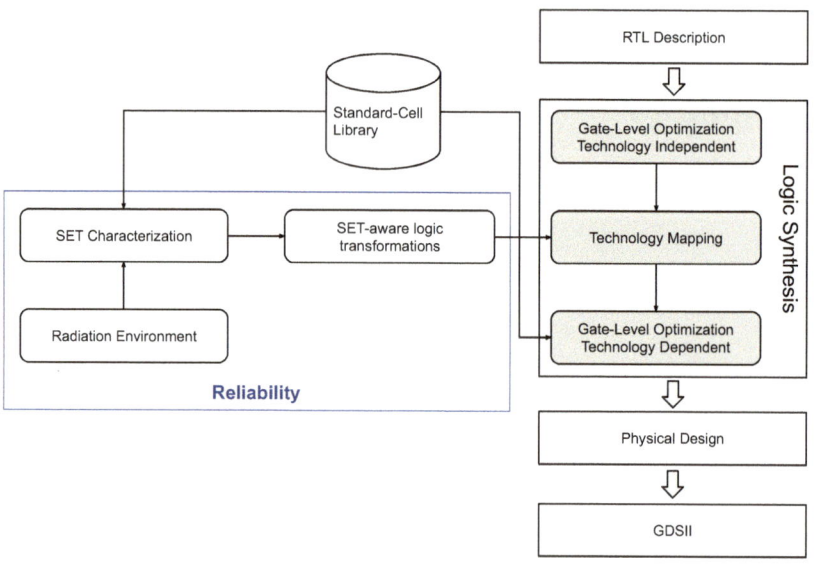

Fig. 6.1 Integration of SET characterization of standard cell libraries for a reliability-driven logic synthesis

Sect. 3.4) can be integrated into the logic synthesis of a cell-based circuit design, as illustrated in Fig. 6.1. Starting from a register transfer level (RTL) description, logic synthesis translates a function into a netlist description of logic gates using a designated standard cell library. The logic synthesis of a VLSI circuit comprises three main processes [11]:

1. **Gate-level optimization independent of technology**: Boolean equations described in the RTL are optimized to minimize size and the number of literals.
2. **Technology mapping**: Transforms each logic function into a logic gate (NAND, NOR, AND, OR, etc.) from the provided cell library.
3. **Technology-dependent gate-level optimization**: Optimizations on the gate netlist are conducted to minimize delay in critical paths, power consumption, and area usage.

RTL, or register transfer level, serves as an abstraction for digital circuits, emphasizing the movement of data between registers. It offers a middle ground between high-level concepts and low-level transistor details.

The gate netlist, a logic-level representation of the circuit, includes gate instances from the standard cell library and their corresponding port connectivity. Consequently, the logic synthesis has a major impact on the resulting gate netlist and thus on the single-event effect (SEE) immunity of the final circuit design. During technology mapping, the technology-independent circuit is broken down into fundamental primitive logic cells (e.g., INV, NAND, or NOR gates). Subsequently, after the decomposition, a pattern-matching process identifies structural and functional patterns to be utilized in the covering process. Here, the optimal patterns are chosen based on a cost function considering factors such as delay, area, and power consumption. Hence, by evaluating the SET immunity of basic logic cells and their combinations, it becomes feasible to develop a reliability-driven cost function and integrate it into the technology mapping process.

The physical design process translates the synthesized gate netlist into the geometric representations used for manufacturing, known as the circuit layout. This stage involves placing each logic cell layout and routing its connections to minimize wire length and optimize power/performance. However, the study conducted by Entrena et al. [3] proposed a cell placement approach focused on reducing charge sharing effects, thereby improving the single-event rate (SER) performance of the circuit.

Similarly, by leveraging the SET characterization methodology on a cell library, a set of SET-aware logic transformations can be integrated into the logic synthesis stage to enhance the SET immunity of the final synthesized gate netlist. Figure 6.2 illustrates the SET cross section of the six most commonly used standard cell logic gates for two-particle linear energy transfer (LET) values.

The AND-OR inverter (AOI) and the OR-AND inverter (OAI) gates implement a larger logic function, resulting in a larger layout area. Consequently, a higher SET cross section is observed when compared with the primitive logic cells. Based on this information, a SET-aware technology mapping could be adopted by assigning a reliability cost to each logic gate. The weight or cost of each gate can be calculated based on the radiation requirements of the mission and the SET cross section or the

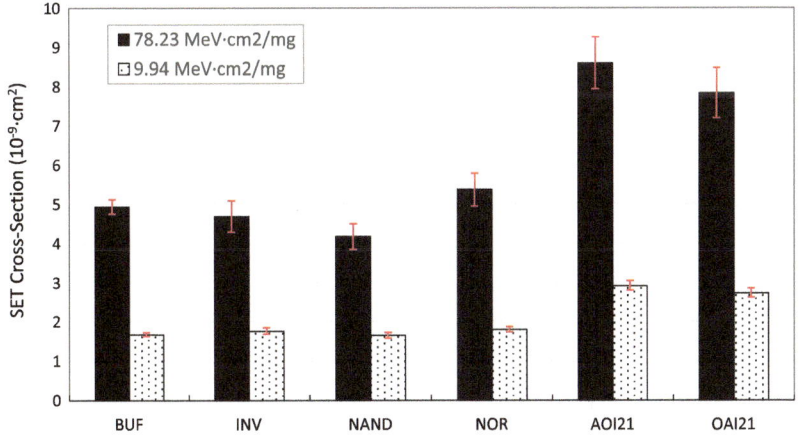

Fig. 6.2 SET cross section for eight standard cell gates from 45 nm NanGate [12], for LET equal to 78.23 and 9.94 MeV.cm^2/mg

estimated in-orbit SET rate. This strategy would enable the selection of gate types during technology mapping with consideration for their impact on circuit reliability in radiation-prone environments.

6.1.1   Multiple V_{th} Cells and Voltage Scaling

A primary objective of logic synthesis is to minimize delay in critical paths. This is achieved by selecting cells with lower propagation times, often facilitated by the adoption of multiple threshold voltage (V_{th}) circuits [13]. Devices with lower V_{th} values exhibit faster switching times, thereby accelerating circuit operation. However, this comes at the expense of increased static power consumption due to higher leakage currents. Conversely, employing high V_{th} devices reduces leakage currents but may lead to performance degradation. Consequently, multiple V_{th} cells are commonly utilized to optimize the gate netlist with respect to both delay and power consumption [14].

An efficient SET characterization methodology can also address the assignment of multiple V_{th} values. Figure 6.3 illustrates the characterization of standard cells using high-performance (HP) process technology, featuring low V_{th} devices, and low-power (LP) process technology, featuring high V_{th} devices. As expected, circuits based on LP technology exhibit a higher overall cross section. This aligns with prior research demonstrating that increased threshold voltage degrades driving strength [15–17]. Notably, NAND gates are the most sensitive to variations in V_{th}, with cross-section increase of 95%. Despite their higher cross sections, complex logic gates such as AOI21 and OAI21 demonstrate relatively low increases.

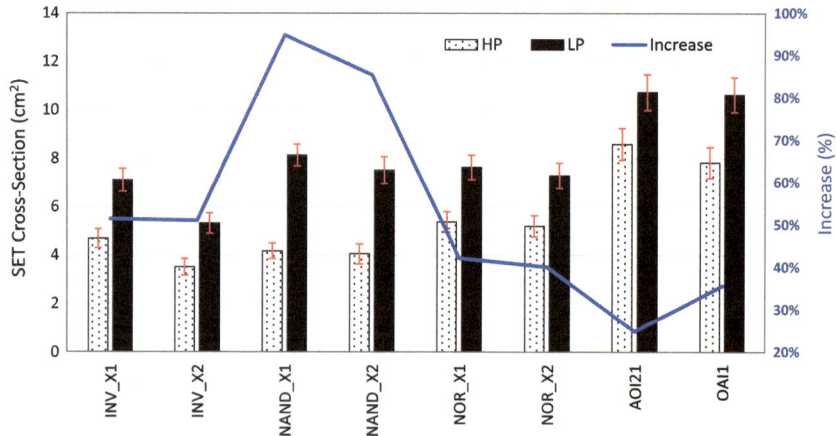

Fig. 6.3 Impact of different threshold voltage devices: High-Performance (HP) vs. Low-Power (LP) devices

Besides impacting the SET cross section of logic gates, the different V_{th} devices can also impact the masking effects and the SEU cross section of memory cells.

While dynamic voltage scaling is another technique for low-power systems [18], reducing supply voltage increases delay and radiation sensitivity [19, 20]. In Fig. 6.4, the SET cross section of each gate is estimated while considering supply voltage scaling from 1 V down to 0.4 V, approaching the near-threshold regime.

Decreasing the supply voltage directly diminishes the critical charge necessary to observe an SEE in the circuit output [19]. Notably, at nominal voltage, NAND gates offer a lower SET cross section compared to NOR gates. However, under voltage scaling to 0.4 V, NOR gates exhibit a lower cross section. This discrepancy stems from the varying impact of drive capability on the transistor networks present in each gate design. Consequently, for low-power systems employing dynamic voltage scaling, logic synthesis should favor NOR gates over NAND gates to enhance radiation hardness.

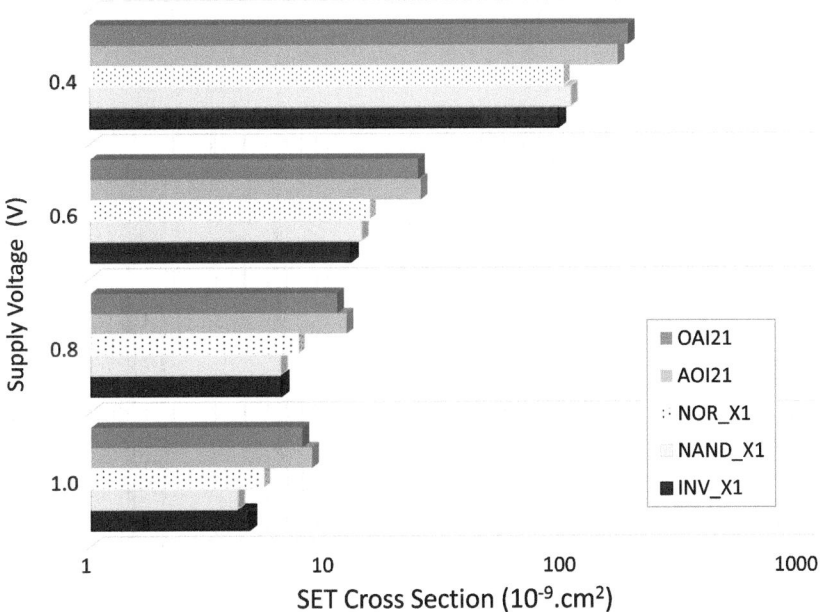

Fig. 6.4 Estimation of the dynamic voltage scaling impact on the SET cross section of the standard cells

6.1.2 Technology Mapping

Reliability-aware logic synthesis, incorporating mitigation strategies for soft errors, involves hardening a complex circuit by selectively employing logic gates that minimize SET generation or propagation in the most vulnerable sub-circuits of a complex VLSI design [5]. For example, the radiation robustness of the circuit can be enhanced by selecting the optimal combination of standard cells that facilitates the pulse quenching effect (PQE) induced by inter-cell charge sharing in electrically related combinational circuits [21, 22]. As previously mentioned, technology mapping is responsible for translating Boolean logic functions described in RTL codes into actual physical logic gates available in a cell library.

Consider the buffer gate, a commonly used circuit amplifier in VLSI circuit design, depicted in Fig. 6.5. Unlike the inverter gate, a digital buffer outputs a signal of the same logic state as its input signal. Buffer insertion, also known as repeater insertion, is a well-established technique in VLSI systems to enhance circuit performance in submicron technology [23]. Moreover, with technology scaling, buffer insertion becomes increasingly important due to the rise in wire delays [24]. Consequently, it is crucial to assess the radiation robustness of buffer gates provided by cell libraries.

From the RTL description of a circuit, the implementation of the buffer Boolean function can be synthesized into either a single buffer (BUF) gate, typically available in the cell library, or two interconnected inverter (INV) gates. Although both implementations realize the same logic function, different radiation sensitivities may be expected due to their distinct layout implementations provided by standard cell design and the cell placement obtained in physical synthesis. Similar differences can be anticipated when implementing the logic functions OR and AND. To analyze which gate combination offers the best SET robustness, in the work developed at [22], the BUF, AND, and OR gates were evaluated under heavy ions and compared with INV, NAND, and NOR gates coupled with an inverter in their output.

Three distinct horizontal cell placement configurations were analyzed to assess the effectiveness of charge sharing and its consequent pulse quenching effect.

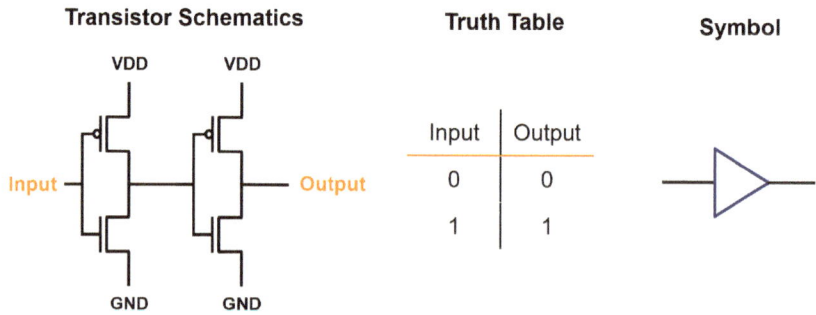

Fig. 6.5 Transistor schematics, truth table, and symbol of a digital CMOS buffer gate

Vertical placements (electrically connected cells in separate rows) were excluded from this study as prior research has shown that they eliminate the pulse quenching effect due to increased nodal separation and the presence of well contacts that significantly reduce charge sharing efficiency [25]. All cells were minimum-sized drive strength (X1) designs from the bulk 45nm technology cell library. The SET cross section (σ_{SET}) for the BUF gate and the connected inverters (with the second inverter placed four minimum cell widths, i.e., 1520 nm from the output of the first inverter) are presented in Fig. 6.6. Clearly, the BUF design has a lower overall SET cross section. This design places the two sensitive nodes closer together compared to the INV + INV setup, enhancing the charge sharing effect. Due to the electrical relationship between these sensitive nodes (inverting stage), pulse quenching effect (PQE) is observed, reducing sensitivity to radiation-induced transient pulses.

Moreover, upon examining the layout design of the buffer gate, it becomes apparent that the first-stage inverter has smaller transistor sizing than the second-stage inverter, as depicted in Fig. 6.7. By reducing the transistor sizing of the first stage, the SET pulse propagated to the second stage is shorter than in the INV+INV design. As the drain collection area in the second inverter remains unchanged in both designs, in the buffer gate, a shorter transient pulse is induced in the first inverter stage while maintaining the same PQE effectiveness in the second-stage inverter, demonstrating its superior robustness.

While the optimal implementation for the buffer boolean function has been shown to be the BUF gate from the studied cell library, for the OR boolean function, the NOR + INV circuit might present a lower SET cross section [22]. Despite the reduced transistor sizing in the internal NOR circuit within the OR gate layout, the NOR+INV circuit continues to offer a lower σ_{SET}, highlighting

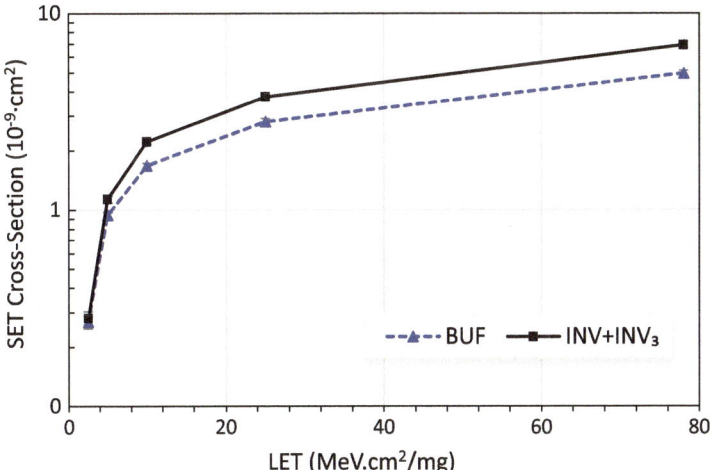

Fig. 6.6 SET cross-section curve for the BUF gate and the INV + INV$_3$ circuit. The inverter gate INV$_3$ is placed 1520 nm from the output of the first inverter

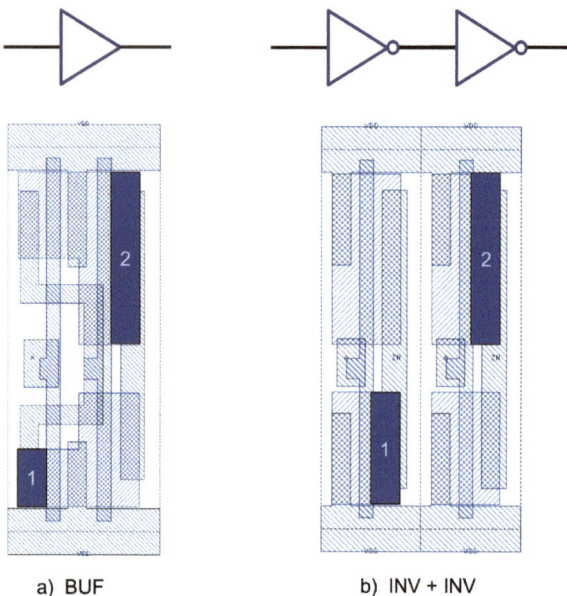

a) BUF   b) INV + INV

Fig. 6.7 Comparison of the collecting drain area of the sensitive electrodes in the first and second stage inverters from the BUF gate and INV + INV design. In the case of the INV + INV$_3$ circuit, INV$_3$ is placed 1520 nm from the output of the first inverter, increasing the distance between the collecting drain areas

its robustness. Unlike the inverter gate, NOR and NAND gates include transistors in series, potentially degrading the recovering current of the circuit. Moreover, while the NAND + INV circuits do not exhibit a lower σ_{SET} compared to the NOR + INV circuits, the NMOS device in the NAND gate's output inverter dominates the pulse quenching effect when the input vector is (1, 1), resulting in a different response compared to the NOR + INV schemes. These findings highlight the importance of evaluating different logic gate combinations during synthesis to achieve optimal radiation hardness for reliable VLSI circuits.

In addition to the basic primitive logic gates (INV, NAND, NOR, etc.), standard cell libraries also offer complex logic gates such as the AND-OR inverter (AOI) and OR-AND inverter (OAI) cells. The usage of such standard cells reduces the number of transistors in the circuit, resulting in denser layouts, decreased power consumption, and smaller area requirements [26]. For instance, the Boolean logic function represented in Eq. 6.1 can be realized using basic logic standard cells like NOR and AND gates, or directly employing the complex gate AOI [27].

$$Y = \neg(A1 \wedge (B1 \vee B2)) \tag{6.1}$$

Implementing Equation 6.1 with the AOI21 gate reduces the transistor count by 40% compared to using an AND gate coupled with a NOR gate. Similarly, the

OAI21 gate achieves the same reduction for the complement of Eq. 6.1. Although the power consumption of complex logic gates is reduced, predicting their radiation sensitivity is challenging. The radiation performance of complex logic gates AOI21 and OAI21 is compared with their corresponding implementations using only basic logic gates such as AND, OR, NAND, and NOR.

The SET cross-section curves considering only P-hit interactions, is shown in Fig. 6.8. In this scenario, all PMOS devices are turned off by setting the input signals to (1, 1, 1). It clearly shows that the implementation containing the basic logic cells AND + NOR provides a lower SET cross section than the AOI21 for the entire LET range. Both circuits present the same threshold LET, whereas there is a SET cross-section difference of approximately a factor of 2. The charge sharing effect and more importantly the logical masking between the AND gate and NOR gate are responsible for this reduced number of observed SET in the output of the AND + NOR implementation [27].

Any SET induced at the AND gate will be filtered by the logic of the NOR gate as observed in the truth table shown in Fig. 6.9. The output of the NOR gate will remain at logic zero as long as the secondary input remains at logic one. This masking effect is not observed for the N-hit configuration. However, by analyzing the structure of the combinational logic and the electrical simulation results, the SETs observed for the AND gate are logically masked by the NOR gate, thus reducing the overall drain sensitive area to the PMOS devices issued in the NOR gate [27].

Fig. 6.10 presents the comparison between the SET cross section of the AND + NOR implementation and the standalone NOR gate. It can be observed that logical masking is effective by reducing the sensitivity of the circuit close to the sensitivity of the standalone NOR gate.

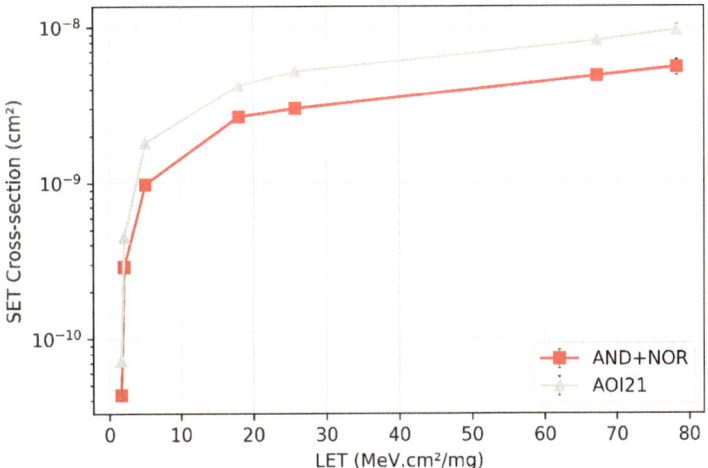

Fig. 6.8 SET cross-section curve of P-hit interactions for the complex logic AOI21 gate and AND + NOR implementation

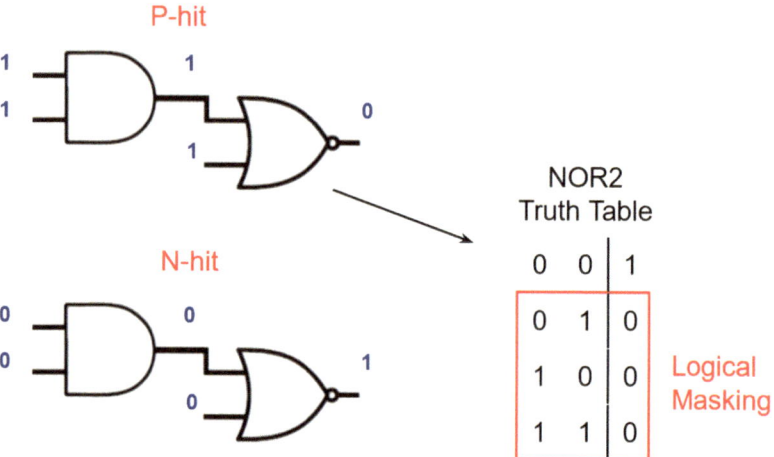

Fig. 6.9 Logical masking effect for the P-hit configuration in the combinational logic circuit AND + NOR. There is no logical masking when both inputs are set to logic zero

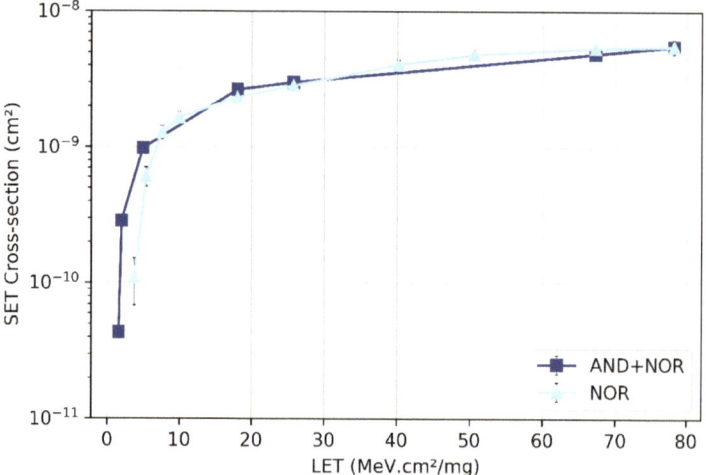

Fig. 6.10 Comparison of SET cross-section curve of the AND + NOR implementation and the standalone NOR gate

Considering only the N-hit interactions, the SET cross sections are approximately the same for high LET ions due to very similar N-hit sensitive areas. Further, there is no contribution of logical masking effect for the input signals considered. As shown in the truth table of the NOR gate in Fig. 6.9, its output is determined whenever one of its input is at logic one. Then, as originally both inputs are set to logic zero, whenever a generated SET at the AND gate propagates to the NOR gate, it will be able to propagate to its output in the case of not being electrically attenuated. In this case, only the electrical masking effect takes place.

6.2 Pin Assignment

The SET characterization of logic gates presents an input dependence due to the varying interplay between sensitive collecting drain areas and restoring current, as previously discussed in Chap. 5. While signal switching activity has long been used to estimate power consumption in VLSI circuit design, it can also support reliability analysis, as demonstrated in [28–30]. Generally, the SET cross section of a digital circuit is given for a specific input signal combination or for the arithmetic mean between the cross section obtained for each input signal combination of the truth table, i.e., the same probability to each input combination is considered. However, the predictive SET characterization (described in Chap. 3) can incorporate signal probability information from a given system application to estimate a more realistic cross section. By considering signal probabilities, more application-efficient mitigation transformations can be proposed in circuit synthesis (Fig. 6.1).

Signal switching activity in VLSI systems serves multiple purposes, including estimating power consumption and performing timing analysis [31–33]. Given the sensitivity of circuits to SET, which is influenced by layout and operational factors such as input signals and internal states, signal probability can be leveraged to enhance circuit reliability [34]. Signal probability-based reliability analysis (SPRA) is an effective tool for analyzing the reliability of VLSI circuits at the gate level [29, 35, 36]. Here, reliability refers to the confidence level that the output will be functionally correct given a fault probability. Initially, SPRA methods focused on physical defects due to wear-out mechanisms or process variability [35]. However, with increasing interest in reducing soft error rates, research has expanded to analyze SET [7, 9, 29, 36]. For instance, [9] proposed a cell placement strategy based on defining bad and good pairs of logic gates to minimize circuit error rates. Similarly, [7] suggested a cell placement approach considering signal probability and its impact on the pulse quenching effect.

In this section, we will discuss how a reliability-driven pin assignment optimization based on the input dependence of SET sensitivity of logic gates can be used to improve the robustness of a given complex digital circuit.

6.2.1 Optimization of Pin Assignment for Single-Event Transients

Pin assignment in logic synthesis enhances power and performance metrics in VLSI circuit design by exploiting the functional equivalence of input pins of logic gates [11, 37, 38]. Consider the NAND gate and its truth table in Fig. 6.11. A symmetric input relationship is evident when both input signals are not identical (i.e., A ≠ B): the output signal is determined whenever one of its inputs is in the low logic level, regardless of the input pin (a or b). The interchangeable input combinations

Fig. 6.11 Schematic of the transistor network of a two-input NAND gate and its truth table. The interchangeable input combinations are highlighted by the red rectangle

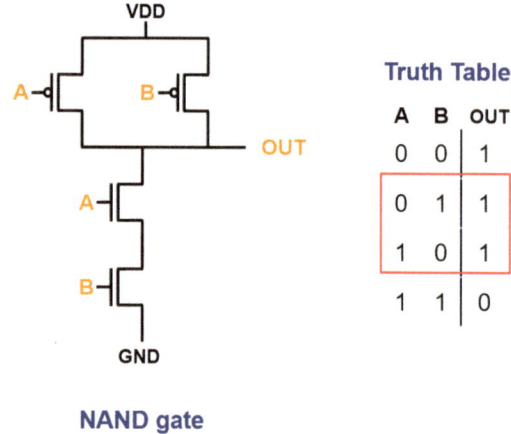

Truth Table

A	B	OUT
0	0	1
0	1	1
1	0	1
1	1	0

NAND gate

are highlighted in Fig. 6.11 (red rectangle). This symmetric input relationship holds true for all two-input basic standard cell gates.

As each input pin of a logic gate presents different electrical characteristics depending on the transistor network, power-driven logic synthesis assigns the input signal with lower switching activity to the pin with higher capacitance. Similarly, a timing-driven optimization can apply pin permutation between symmetric input pins such that the late-arriving signal is always connected to the input pin with the lowest intrinsic delay [11]. This process is known as rewiring or pin swapping [37, 38]. Considering that cross section of logic gates is dependent on the input stimuli, a reliability-aware synthesis can be proposed based on the cell symmetric inputs and signal probabilities to improve the vulnerability of the circuit through optimal pin assignment.

The proposed SET-aware pin assignment optimization in a cell-based circuit design flow is illustrated in Fig. 6.12. Beginning with a circuit description in RTL, logic synthesis optimizes each Boolean function and maps it to logic gates available in the standard cell library. The resulting gate netlist, typically optimized for timing, area, and/or power, is then utilized alongside the standard cell library for the SET-aware pin assignment optimization.

The first step involves the input-based SET characterization, where the symmetric input relationship of each standard cell is identified, and the SET cross section for interchangeable input combinations is calculated. For example, the cross section for input combinations (0, 1) and (1, 0) is determined. Based on the SET characterization results, a set of pin assignment rules is defined. This entails assigning an input pin for each standard cell so that the net with the lowest signal probability is connected, ensuring that the most sensitive interchangeable input combination has the lowest probability of occurrence.

In the second step, considering the switching activity of the primary inputs, the calculation of signal probability can be performed for internal net connections, as depicted in Fig. 6.13. Using the Boolean function of each gate, an equation is derived to calculate the probability of the output signal being at logic value 1, denoted

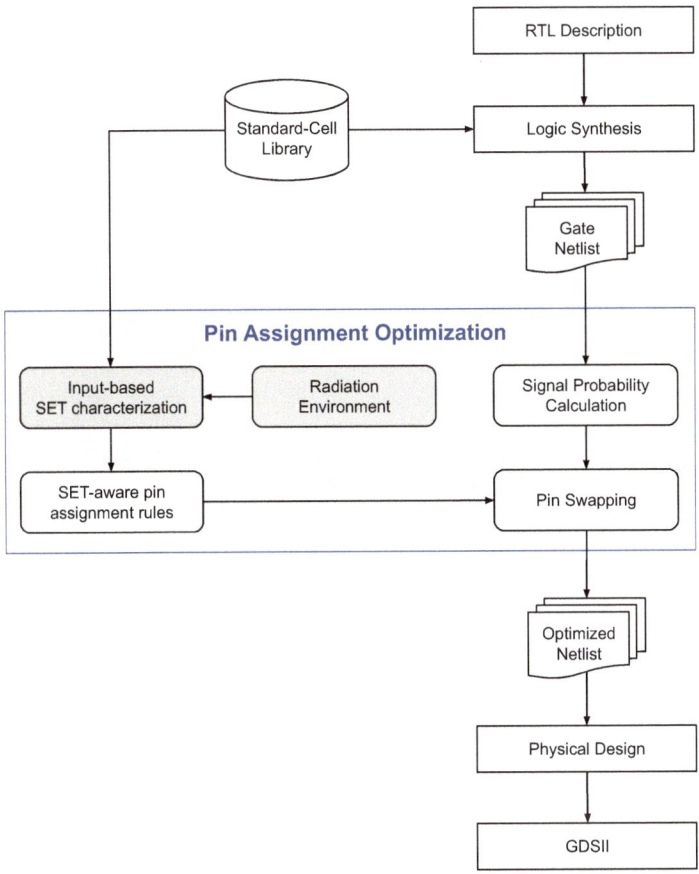

Fig. 6.12 SET-aware pin assignment optimization in a cell-based VLSI circuit design flow

as $P_{GATE}(output = 1)$. In this study, the Parker-McCluskey method [39] was employed, assuming uncorrelated primary inputs with equal switching activity of 50% (signal probability equals 0.5). Although temporal and spatial correlations are not considered in our analysis, more sophisticated and accurate signal probability estimation methods can be seamlessly integrated into the development flow depicted in Fig. 6.12. Utilizing basic probability theory, the output signal probability equations for the logic gates used in the combinational circuit in Fig. 6.13 are presented in Table 6.1.

For instance, consider the inverter gate. Given the input probability $p(a = 1) = 1$, i.e., the signal is always at logic value 1, considering its Boolean function, the probability of the output signal to be at logic value 1 is 0 ($P_{INV}(output = 1) = 0$). Thus, the signal probability equation for the inverter can be expressed by Eq. 6.2:

$$P_{INV}(\text{output} = 1) = 1 - p(a = 1) \qquad (6.2)$$

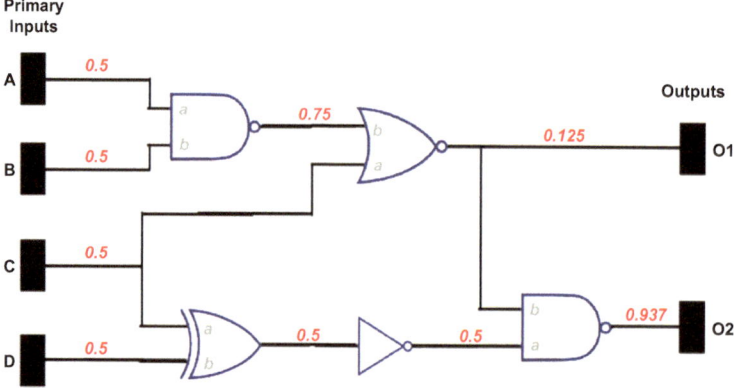

Fig. 6.13 Signal probability estimation for a combinational logic circuit. Primary input signal probability is assigned to 0.5

Table 6.1 Signal probability estimation for the INV, NAND, NOR, and XOR gates

Cells	Input signals	Output signal probability[a]
INV	1: a	$P_{INV} = 1 - p_A$
NAND	2: a, b	$P_{NAND} = 1 - (p_A \times p_B)$
NOR	2: a, b	$P_{NOR} = (1 - p_A) \times (1 - p_B)$
XOR	2: a, b	$P_{XOR} = p_A \times (1 - p_B) + p_B \times (1 - p_A)$

[a] Signal probability is the probability of the signal to be at logic value 1

For clarity, the probabilities $P_{GATE}(output = 1)$, $p(a = 1)$, and $p(b = 1)$ will be shortened to P_{GATE}, p_A, and p_B. Following the signal probability calculation step in the Fig. 6.12 is the pin swapping process. In this process, with the pin assignment rules and the signal probabilities, an optimization algorithm can identify which pin should be swapped in order to reduce the occurrence of the interchangeable input combination with the higher SET cross section. After the pin swapping, the standard design flow is performed with the optimized netlist.

In order to obtain important reliability information to be addressed in the optimization process, the SET characterization methodology is aligned to the identification of the input dependence [34, 40]. The sensitivity of each standard cell is extracted from the layout design in the Graphical Design System (GSDII) file. Thus, the Monte Carlo simulation tool, MC-Oracle [41], is used to obtain the SET currents. In order to consider the input signal probabilities, Eq. 6.3 is used. Given n input combinations, the overall gate SET cross-section σ_{Gate} can be estimated from the input SET cross-section $\sigma_{SET}(i)$ and the input probability $p(i)$:

$$\sigma_{Gate} = \sum_{i=0}^{n} \sigma_{SET}(i) \times p(i) \tag{6.3}$$

The input cross-section $\sigma_{SET}(i)$ is provided by the SET characterization, while the input probabilities are provided by the signal probability calculation in Fig. 6.13. Then, this equation is used in the Pin Swapping process to evaluate when the input pins assigned from the logic synthesis should be swapped to decrease the gate SET cross section. Similarly, this process can also adopt the soft-error rate estimation for a given mission orbit as shown in [40].

6.2.2 Impact on the SET Cross Section of Standard Cells

Based on the cross section of the interchangeable input combinations, an optimized logic synthesis should prioritize the pin assignment of the lowest signal probability in such a way the most sensitive interchangeable input combination obtains the lowest probability of occurrence [40]. The input SET cross section for NAND, NOR, and XOR gates under particle linear energy transfer (LET) of 5 MeV.cm^2/mg is shown in Fig. 6.14. It is possible to identify the most sensitive input combinations for each gate. Considering the interchangeable input, i.e., (0, 1) and (1, 0), the NAND gate is the only cell to show the lowest cross section for (1, 0), while the NOR and XOR gates present (0, 1). It implies that, considering low-particle LET, the lowest signal probability should be assigned to the input B for the NAND gate and to the input A for the NOR and XOR gates.

When adopting a three-input logic function such as AOI21 and OAI21, it is not possible to obtain a complete symmetric input relationship as observed for the two-input gates. It is necessary to identify the interchangeable input combinations and also what we denominate as the controllable input pin, i.e., the pin that controls the

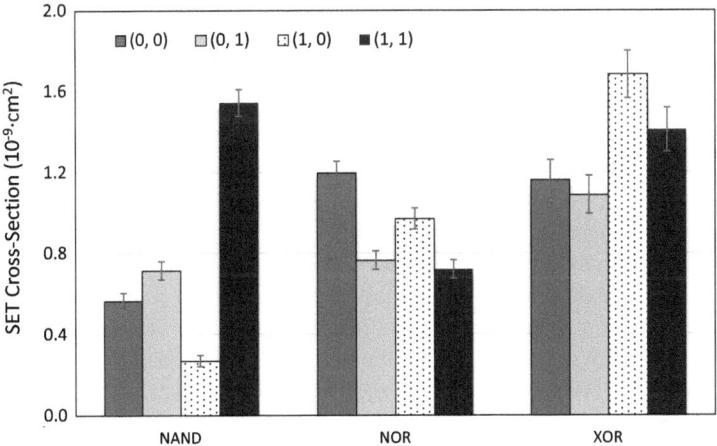

Fig. 6.14 Input SET cross section for the NAND, NOR, and XOR gates under a particle with LET $= 5$ MeV.cm^2/mg

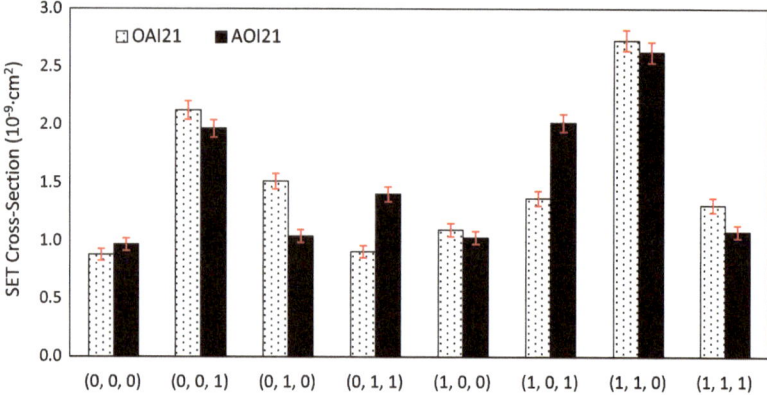

Fig. 6.15 Input SET cross section for the AOI21 and OAI21 gates under a particle with LET = 5 MeV.cm^2/mg

output of the function and the pin assignment cannot optimize it; otherwise, it will interfere with the correct logic function of the circuit. In [40], this methodology is explained in detail and the results for the AOI21 and OAI21 under an LET of 5 MeV.cm2/mg is shown in Fig. 6.15.

By applying Eq. 6.3, we can obtain the overall gate SET cross-section σ_{Gate} for a given input scenario. Figure 6.16 provides the gate SET cross-section σ_{Gate} curves for the NOR gate considering two signal probability scenarios [a: 0.9, b: 0.1] and [a: 0.1, b: 0.9]. For high LET, a slight reduction is observed in the cross section when the lowest signal probability is assigned to the input pin b as it reduces the impact from the input combination (0, 1). However, for low LET, the cross section can be drastically reduced if the lowest signal probability is assigned to the input pin a, instead. With the reduction of the particle LET, the impact on the overall gate sensitivity is dominated by the input combination (1, 0), being comparable to the worst-case input scenario for this logic gate, the input (0, 0). For instance, a SET cross-section reduction of solely 9% can be obtained for the pin assignment [a: 0.9, b: 0.1] when the LET is 78.23 MeV.cm^2/mg, while a reduction up to 86% is expected for [a: 0.1, b: 0.9] under 2.53 MeV.cm^2/mg.

To verify this impact within a mission environment, the gate reliability can be examined in terms of in-orbit SET rates. Figures 6.17 and 6.18 present the in-orbit SET rates calculated using the OMERE tool [42], based on NOR gate SET cross-section curves shown in Fig. 6.16. The standard method for calculating the SEE rate, as specified by the European Cooperation for Space Standardization (ECSS), employs the integral rectangular parallelepiped (IRPP) method [43]. The SEE rate is derived through the convolution of the cross-section data with the particle flux in the mission orbit.

In this study, the SET rates were computed for both a geostationary orbit (GEO) and a low-Earth orbit (LEO) International Space Station (ISS) orbit. Employing a fixed shielding of 1 g/cm^2, the international standard ISO 15390 model is utilized for

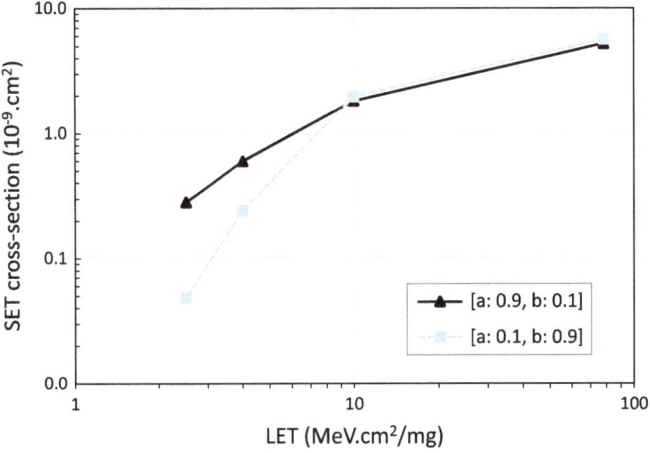

Fig. 6.16 SET cross-section curves for the two-input NOR gate considering two signal probability scenarios: Lowest signal probability assigned to input pin a [a:0.1, b:0.9]; lowest signal probability assigned to input pin b [a:0.9, b:0.1]

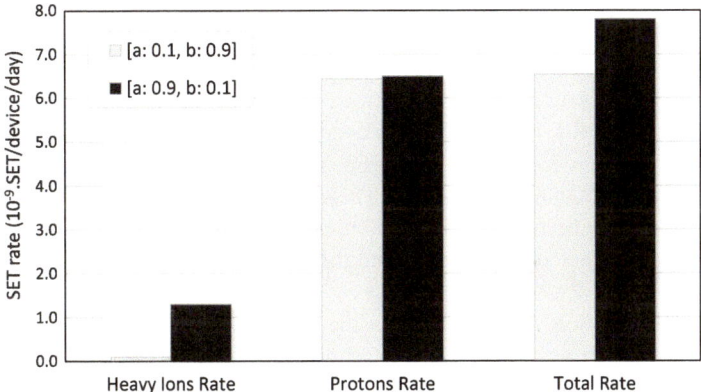

Fig. 6.17 In-orbit SET rate calculated based on the proposed SET characterization considering the International Space Station (ISS) orbit, 400 km, 51.64°

the galactic cosmic rays (GCR) [44], while the NASA AP8MIN trapped radiation model is adopted for the trapped proton flux under solar minimum conditions [45]. The results are segmented into heavy-ion, proton, and total rates (combining heavy-ion and proton rates).

The significant impact of the pin assignment is evident in heavy-ion rates, showcasing a reduction of approximately 83% and 92% on the SET rate for the GEO and ISS orbits, respectively. However, in the ISS orbit, protons are anticipated to dominate the SEE rate, as depicted in Fig. 6.17. By accounting for the contributions

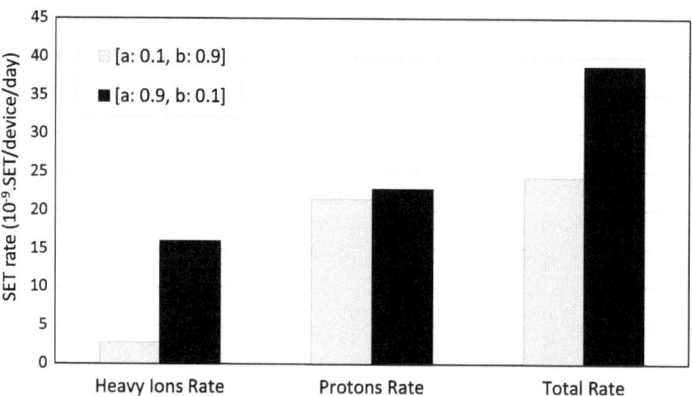

Fig. 6.18 In-orbit SET rate calculated based on the proposed SET characterization considering a Geostationary Orbit (GEO), 35,784 km

of heavy ions and protons to the SET rate of the two-input NOR gate, a SET-aware pin assignment could yield a reduction of 37% and 16% in its total SET rate for the GEO and ISS orbits, respectively.

This analysis can be extended to encompass every gate within a cell library or a complex circuit. Such an approach facilitates the development of customized rules-based optimization and a SET-aware pin assignment during logic synthesis. For instance, in [40], this technique was applied in arithmetic benchmark circuits and up to 28% reduction on the SET rate was observed, depending on the circuit architecture and mission orbit. This optimization technique provides no area overhead and can be used along with other hardening techniques. Additionally, it does not impact the cell placement, and the net routing is impacted minimally and locally.

This optimization technique holds the potential to significantly enhance the benefits derived from the hardening techniques discussed earlier in this book, as it has been shown that the efficacy of layout and circuit-level techniques often exhibits input dependence. Consequently, pin assignment can be employed to maximize their efficiency. For example, this approach has been verified for the triple modular redundancy (TMR) optimization strategies discussed in [46]. In this work, four majority voter architectures were studied, and signal probability analysis was used to optimize TMR block insertion. While complex gate voters offered a smaller layout footprint, they exhibited a higher overall SET cross section due to increased charge collection area from upsized transistors. Conversely, basic logic gate architectures displayed a higher dependence on input signal probabilities due to logical masking effects. These findings suggest that incorporating an input dependence analysis into SET assessment can guide selective TMR block insertion in critical circuit nodes based on their signal probability distribution. This approach can optimize radiation hardening while minimizing area overhead and others.

6.3 Summary

In this chapter, the evaluation of circuit-level techniques is discussed, emphasizing the significance of reliability-driven synthesis in the VLSI design flow. The applicability of a detailed SEE characterization to support synthesis algorithms targeting SEE resilience is discussed. In this sense, a cell-based characterization can provide insights on the sensitivity of logic gates considering various device technologies (e.g., high performance vs. low power) and operating conditions (e.g., dynamic voltage scaling, input signal probability, etc.).

It was shown that the technology mapping process has a crucial role in determining the overall robustness of complex circuits. Since a single Boolean logic function can be implemented in various ways using different logic gates, logic synthesis in ASIC designs can benefit from incorporating logic transformations focused on reliability cost functions to mitigate the occurrence of transient pulses. For instance, the usage of complex logic gates might reduce the logical masking capabilities of a given combinational circuit.

Given the input dependence of the SEE sensitivity of logic circuits, an optimization methodology is proposed to improve the overall circuit hardness through pin swapping based on the signal probability. As shown in Chap. 5, the input dependence is attributed to the different driving capabilities and the influence of the layout design on the SET robustness. Given that the influence of layout design varies with LET, the relationship between input signal and gate SET cross section is also demonstrated to be LET dependent. The impact of adopting signal probability evaluation and pin assignment has shown the greatest cross-section reduction for low-particle LET. Considering the two-input gates, the signal with the lowest probability should be assigned to the input B for the NAND gate and to the input A for the NOR and XOR gates. For the three-input gates, different conclusions can be drawn based on the signal probability of the controllable input.

Pin assignment alone has the potential to reduce the SET rate, but it can also serve as an optimization technique to maximize the effectiveness of other layout and circuit hardening techniques. For example, a selective TMR block insertion can be developed based on the signal probability of the critical nodes and the SET cross section of several majority voter designs. The versatility of this optimization technique, grounded in input signal probability and input dependence analysis, offers a multitude of possibilities for achieving a more reliable and robust electronic system.

Highlights

- The logical and physical synthesis processes of a VLSI circuit play a significant role in determining the performance characteristics of a VLSI circuit, from power, timing, and area consumption to the reliability against soft errors.

(continued)

- A Boolean logic function can be implemented in several ways during the technology mapping process, and therefore, the overall SEE susceptibility will be different for each design version.
- Using complex logic gates can reduce the logical masking capabilities when compared with their counterpart design version using basic logic gates such as NAND, NOR, and INV gates.
- The SET input dependence of logic gates can be used to optimize the overall circuit reliability when the signal probability of its input is considered.
- Reliability-aware pin assignment can improve the overall robustness of a circuit on its own and, even more significantly, maximize the efficiency of several other hardening techniques without introducing any area overhead.

References

1. Ian Kuon and Jonathan Rose. Measuring the gap between FPGAS and ASICS. *IEEE Transactions on computer-aided design of integrated circuits and systems*, 26 (2): 203–215, 2007.
2. Paul S Zuchowski, Christopher B Reynolds, Richard J Grupp, Shelly G Davis, Brendan Cremen, and Bill Troxel. A hybrid ASIC and FPGA architecture. In *IEEE/ACM International Conference on Computer Aided Design, 2002. ICCAD 2002.*, pages 187–194. IEEE, 2002.
3. L. Entrena, A. Lindoso, E. S. Millan, S. Pagliarini, F. Almeida, and F. Kastensmidt. Constrained placement methodology for reducing SER under single-event-induced charge sharing effects. *IEEE Transactions on Nuclear Science*, 59 (4): 811–817, Aug 2012. ISSN 0018-9499. doi: https://doi.org/10.1109/TNS.2012.2191796.
4. Cristiano Lazzari, Gilson Wirth, Fernanda Lima Kastensmidt, Lorena Anghel, and Ricardo Augusto da Luz Reis. Asymmetric transistor sizing targeting radiation-hardened circuits. *Electrical Engineering (Archiv fur Elektrotechnik)*, 94 (1): 11–18, 2012.
5. Daniel B Limbrick, Nihaar N Mahatme, William H Robinson, and Bharat L Bhuva. Reliability-aware synthesis of combinational logic with minimal performance penalty. *IEEE Transactions on nuclear science*, 60 (4): 2776–2781, 2013.
6. Bradley T Kiddie and William H Robinson. Alternative standard cell placement strategies for single-event multiple-transient mitigation. In *2014 IEEE Computer Society Annual Symposium on VLSI*, pages 589–594. IEEE, 2014.
7. Yankang Du, Shuming Chen, and Biwei Liu. A constrained layout placement approach to enhance pulse quenching effect in large combinational circuits. *IEEE Transactions on Device and Materials Reliability*, 14 (1): 268–274, 2013.
8. Aiman H El-Maleh and Khaled AK Daud. Simulation-based method for synthesizing soft error tolerant combinational circuits. *IEEE Transactions on Reliability*, 64 (3): 935–948, 2015.
9. Mohamad Imran Bandan, Samuel Pagliarini, Jimson Mathew, and Dhiraj Pradhan. Improved multiple faults-aware placement strategy: Reducing the overheads and error rates in digital circuits. *IEEE Transactions on Reliability*, 66 (1): 233–244, 2017.

10. T. Lange, A. Balakrishnan, M. Glorieux, D. Alexandrescu, and L. Sterpone. Machine learning clustering techniques for selective mitigation of critical design features. In *2020 IEEE 26th International Symposium on On-Line Testing and Robust System Design (IOLTS)*, pages 1–7, 2020.

11. Soha Hassoun and Tsutomu Sasao. *Logic synthesis and verification*, volume 654. Springer Science & Business Media, 2012.

12. James E Stine, Ivan Castellanos, Michael Wood, Jeff Henson, Fred Love, W Rhett Davis, Paul D Franzon, Michael Bucher, Sunil Basavarajaiah, Julie Oh, et al. FreePDK: An open-source variation-aware design kit. In *2007 IEEE international conference on Microelectronic Systems Education (MSE'07)*, pages 173–174. IEEE, 2007.

13. James T Kao and Anantha P Chandrakasan. Dual-threshold voltage techniques for low-power digital circuits. *IEEE Journal of Solid-state circuits*, 35 (7): 1009–1018, 2000.

14. Guilherme Flach, Tiago Reimann, Gracieli Posser, Marcelo Johann, and Ricardo Reis. Effective method for simultaneous gate sizing and v th assignment using lagrangian relaxation. *IEEE transactions on computer-aided design of integrated circuits and systems*, 33 (4): 546–557, 2014.

15. Y. Q. Aguiar, F. L. Kastensmidt, C. Meinhardt, and R. Reis. Implications of work-function fluctuation on radiation robustness of finfet XOR circuits. In *Radiation Effects on Components and Systems (RADECS) Conference*. IEEE, 2017.

16. Hangfang Zhang, Hui Jiang, Thiago R Assis, Nihaar N Mahatme, Balaji Narasimham, Lloyd W Massengill, Bharat L Bhuva, Shi-Jie Wen, and Richard Wong. Effects of threshold voltage variations on single-event upset response of sequential circuits at advanced technology nodes. *IEEE Transactions on Nuclear Science*, 64 (1): 457–463, 2016.

17. RC Harrington, JA Maharrey, JS Kauppila, P Nsengiyumva, DR Ball, EX Zhang, BL Bhuva, and LW Massengill. Effect of transistor variants on single-event transients at the 14-/16-nm bulk finfet technology generation. *IEEE Transactions on Nuclear Science*, 65 (8): 1807–1813, 2018.

18. Thomas D Burd, Trevor A Pering, Anthony J Stratakos, and Robert W Brodersen. A dynamic voltage scaled microprocessor system. *IEEE Journal of solid-state circuits*, 35 (11): 1571–1580, 2000.

19. Tino Heijmen, Damien Giot, and Philippe Roche. Factors that impact the critical charge of memory elements. In *12th IEEE International On-Line Testing Symposium (IOLTS'06)*, pages 6–pp. IEEE, 2006.

20. Fernanda Lima Kastensmidt, Jorge Tonfat, Thiago Both, Paolo Rech, Gilson Wirth, Ricardo Reis, Florent Bruguier, Pascal Benoit, Lionel Torres, and Christopher Frost. Voltage scaling and aging effects on soft error rate in SRAM-based FPGAs. *Microelectronics Reliability*, 54 (9-10): 2344–2348, 2014.

21. J. R. Ahlbin, M. J. Gadlage, D. R. Ball, A. W. Witulski, B. L. Bhuva, R. A. Reed, G. Vizkelethy, and L. W. Massengill. The effect of layout topology on single-event transient pulse quenching in a 65 nm bulk CMOS process. *IEEE Transactions on Nuclear Science*, 57 (6): 3380–3385, Dec 2010. ISSN 0018-9499. doi: https://doi.org/10.1109/TNS.2010.2085449.

22. Ygor Q. Aguiar, Frédéric Wrobel, J-L Autran, Paul Leroux, Frédéric Saigné, Antoine D Touboul, and Vincent Pouget. Analysis of the charge sharing effect in the SET sensitivity of bulk 45 nm standard cell layouts under heavy ions. *Microelectronics Reliability*, 88: 920–924, 2018.

23. Lukas PPP Van Ginneken. Buffer placement in distributed RC-tree networks for minimal elmore delay. In *IEEE International Symposium on Circuits and Systems*, pages 865–868. IEEE, 1990.

24. Prashant Saxena, Noel Menezes, Pasquale Cocchini, and Desmond A Kirkpatrick. Repeater scaling and its impact on CAD. *IEEE Transactions on Computer-Aided Design of Integrated Circuits and Systems*, 23 (4): 451–463, 2006.

25. He Yibai, Chen Shuming, Chen Jianjun, Chi Yaqing, Liang Bin, Liu Biwei, Qin Junrui, Du Yankang, and Huang Pengcheng. Impact of circuit placement on single event transients in 65 nm bulk CMOS technology. *IEEE Transactions on Nuclear Science*, 59 (6): 2772–2777, 2012.
26. Ricardo Reis. Power consumption & reliability in nanoCMOS. In *2011 11th IEEE International Conference on Nanotechnology*, pages 711–714. IEEE, 2011.
27. Y. Q. Aguiar, Frédéric Wrobel, J-L Autran, Paul Leroux, Frédéric Saigné, AD Touboul, and Vincent Pouget. Impact of complex-logic cell layout on the single-event transient sensitivity. *IEEE Transactions on Nuclear Science*, 66 (7): 1465–1472, 2019.
28. Farid N Najm. Improved estimation of the switching activity for reliability prediction in VLSI circuits. In *Proceedings of IEEE Custom Integrated Circuits Conference-CICC'94*, pages 429–432. IEEE, 1994.
29. Denis Teixeira Franco, Mai Correia Vasconcelos, Lirida Naviner, and Jean-François Naviner. Reliability analysis of logic circuits based on signal probability. In *2008 15th IEEE International Conference on Electronics, Circuits and Systems*, pages 670–673. IEEE, 2008a.
30. Rafael Schvittz, Denis T Franco, Leomar S Rosa, and Paulo F Butzen. Probabilistic method for reliability estimation of SP-networks considering single event transient faults. In *2018 25th IEEE International Conference on Electronics, Circuits and Systems (ICECS)*, pages 357–360. IEEE, 2018.
31. Abhijit Ghosh, Srinivas Devadas, Kurt Keutzer, and Jacob White. Estimation of average switching activity in combinational and sequential circuits. In *DAC*, volume 29, pages 253–269, 1992.
32. Massoud Pedram. Power minimization in IC design: Principles and applications. *ACM Transactions on Design Automation of Electronic Systems (TODAES)*, 1 (1): 3–56, 1996.
33. Bao Liu. Signal probability based statistical timing analysis. In *Proceedings of the conference on Design, automation and test in Europe*, pages 562–567, 2008.
34. Ygor Q. Aguiar, Frédéric Wrobel, Jean-Luc Autran, Paul Leroux, Frédéric Saigné, Vincent Pouget, and Antoine D Touboul. Mitigation and predictive assessment of set immunity of digital logic circuits for space missions. *Aerospace*, 7 (2): 12, 2020a.
35. Sreejit Chakravarty and Harry B Hunt. On computing signal probability and detection probability of stuck-at faults. *IEEE Transactions on Computers*, 39 (11): 1369–1377, 1990.
36. Denis Teixeira Franco, Maí Correia Vasconcelos, Lirida Naviner, and Jean-François Naviner. Signal probability for reliability evaluation of logic circuits. *Microelectronics Reliability*, 48 (8-9): 1586–1591, 2008b.
37. Sunil P Khatri and Kanupriya Gulati. *Advanced Techniques in Logic Synthesis, Optimizations and Applications*. Springer, 2011.
38. André Inácio Reis and Rolf Drechsler. *Advanced Logic Synthesis*. Springer, 2018.
39. Kenneth P. Parker and Edward J. McCluskey. Probabilistic treatment of general combinational networks. *IEEE Transactions on Computers*, 100 (6): 668–670, 1975.
40. Y. Q. Aguiar, F. Wrobel, J.-L. Autran, P. Leroux, F. Saigné, V. Pouget, and A. D. Touboul. Reliability-driven pin assignment optimization to improve in-orbit soft-error rate. *Microelectronics Reliability*, 114: 113885, 2020b.
41. Frédéric Wrobel and Frédéric Saigné. MC-ORACLE: A tool for predicting soft error rate. *Computer Physics Communications*, 182 (2): 317–321, 2011.
42. The OMERE 5.3 software by TRAD and CNES. URL http://www.trad.fr/en/space/omere-software.
43. ECSS Secretariat. Space engineering: Calculation of radiation and its effects and margin policy handbook - ECSS-E-HB-10-12A. 2010.
44. Aircraft Technical Committee ISO/TC 20, Space systems space vehicles, Subcommittee SC 14, and operations. ISO-15390: 2004. Space environment (natural and artificial)-galactic cosmic ray model, 2004.

45. Donald M Sawyer and James I Vette. AP-8 trapped proton environment for solar maximum and solar minimum. Technical report, National Aeronautics and Space Administration, 1976.
46. Y. Q. Aguiar, F. Wrobel, J.-L. Autran, P. Leroux, F. Saigné, V. Pouget, and A. D. Touboul. Design exploration of majority voter architectures based on the signal probability for TMR strategy optimization in space applications. *Microelectronics Reliability*, 114: 113877, 2020c.

Index

A
Asymmetric designs, 99, 101–107, 109

B
Beam interaction, 10, 25
Bulk technology, 66, 67, 75, 96, 121

C
Charge sharing, 23–25, 34, 35, 44, 52, 54, 55,
    57–59, 67, 75, 81, 86, 87, 99, 107, 108,
    117, 120, 121, 123
Chord-length model, 50
Circuit designs, 29, 33, 38, 41, 55, 57, 58, 63,
    70, 74, 94–96, 115–117, 120, 125
Complex-logic gates, 37, 122, 123, 133, 134
Coulombic interactions, 13, 14, 56
Critical charge, 19, 23, 32, 33, 39, 44, 51, 53,
    119

D
Detailed history of recoiling ions induced by
    nucleons (DHORIN) code, 56
DHORIN code, see Detailed history of
    recoiling ions induced by nucleons
    (DHORIN) code
DICE, see Dual interlocked storage cell
    (DICE)
Diffusion, 21–25, 34, 42, 50, 52–59, 67, 69,
    82, 96, 97, 106, 107
Diffusion splitting (DS), 94–110
Drift, 21–23, 25, 52–53, 56–58, 67
Drift-diffusion collection model, 50, 52, 57, 58

DS, see Diffusion splitting (DS)
Dual interlocked storage cell (DICE), 74, 75

E
Electrical masking, 37–39, 94, 105, 109, 124
ELT, see Enclosed layout transistors (ELT)
Enclosed layout transistors (ELT), 68, 69
Event cross-section, 33

F
FDSOI technology, see Fully-depleted SOI
    (FDSOI) technology
Fully-depleted SOI (FDSOI) technology, 67

G
Galactic cosmic rays (GCRs), 1–6, 25, 106,
    131
Gate finger, 97
Gate-level optimization, 116
Gate sizing (GS), 72, 82–96, 99, 109, 110, 115
GCRs, see Galactic cosmic rays (GCRs)
GS, see Gate sizing (GS)
Guard-rings, 69, 70, 96

H
Hadrons, 8, 9, 11, 12, 25
High-performance (HP) process technology,
    118
Holding voltage, 41, 44
Hold time, 39

© The Editor(s) (if applicable) and The Author(s) 2025
Y. Quadros de Aguiar et al., *Single-Event Effects, from Space to Accelerator
Environments*, https://doi.org/10.1007/978-3-031-71723-9

I

In-orbit SET rate, 96, 106–107, 109, 118,
  130–132
Input dependence, 99, 104, 109, 125, 128,
  132–134
Integral rectangular paralleliped (IRPP)
  analytical model, 52, 59, 106, 130
Ionization, 13, 15–21, 25, 32, 36, 50–52,
  54–56, 87
IRPP analytical model, *see* Integral rectangular
  paralleliped (IRPP) analytical model

L

Latching-window masking, 37, 39–40
Layout design, 43, 55, 56, 58, 70, 71, 75, 91,
  97, 98, 108, 110, 115, 121, 128, 133
Layout design through error-aware transistor
  positioning (LEAP) layout technique,
  71, 72, 75
LEAP layout technique, *see* Layout design
  through error-aware transistor
  positioning (LEAP) layout technique
LET, *see* Linear energy transfer (LET)
Linear energy transfer (LET), 16, 33, 43, 44,
  51, 53, 71, 89–91, 98–105, 108–110,
  117, 123, 124, 129, 130, 133
Logical masking, 37–38, 123, 124, 132–134
Logical synthesis, 115, 133
Logic transformations, 117, 133
Low-power (LP) process technology, 118
LP process technology, *see* Low-power (LP)
  process technology
Luminosity, 10, 11

M

Masking effects, 30, 38–40, 43, 44, 87, 94,
  109, 119, 123, 124, 132
MC-Oracle, 49, 50, 55–58, 96, 106, 128
MC simulation, *see* Monte Carlo (MC)
  simulation
Monte Carlo (MC) simulation, 12, 49, 50, 55,
  96, 128
Multiple-node collection, 52, 54, 57, 76

N

Nuclear reactions, 13, 15, 17, 29, 56

P

Parasitic bipolar amplification (PBA), 52, 55,
  67, 76

Partially-depleted SOI (PDSOI) technology,
  67, 68
Particle fluence, 3, 33
Particle flux, 3, 4, 8, 25, 33, 106, 132
PBA, *see* Parasitic bipolar amplification (PBA)
PDSOI technology, *see* Partially-depleted SOI
  (PDSOI) technology
Physical layout techniques, 63
Physical synthesis, 81, 120, 133
Pin assignment, 125–134
Pin swapping technique, 126–129, 133
PIPB, *see* Propagation-induced pulse
  broadening (PIPB)
PQE, *see* Pulse quenching effect (PQE)
Propagation-induced pulse broadening (PIPB),
  39, 96
Pulse quenching effect (PQE), 23–24, 57,
  120–121, 125
PVT variability, 106

R

Radiation hardening, 30, 58, 63–77, 86, 88,
  104, 132
Radiation hardening by design (RHBD)
  techniques, 58, 63, 64, 68–77, 81–110,
  115–134
Radiation hardening by process (RHBP)
  techniques, 63, 65–68, 76, 77
Rectangular parallelepiped (RPP) analytical
  model, 50–52, 58, 59, 130
Register transfer level (RTL), 81, 116, 120, 126
RHBD techniques, *see* Radiation hardening by
  design (RHBD) techniques
RHBP techniques, *see* Radiation hardening by
  process (RHBP) techniques
RPP analytical model, *see* Rectangular
  parallelepiped (RPP) analytical model
RTL, *see* Register transfer level (RTL)

S

SEFI, *see* Single-event functional interruption
  (SEFI)
SEL, *see* Single-event latchup (SEL)
Sensitive volume, 19, 50–52, 58, 66, 67
SEPs, *see* Solar energetic particles (SEPs)
SET, *see* Single-event transient (SET)
SET cross-section, 42, 86, 89–94, 97–110, 117,
  119, 121, 123–126, 128–133
Setup time, 39
SEU, *see* Single-event upset (SEU)
Shallow trench isolation (STI) oxide, 69, 86

Signal probability, 109, 125–134
Silicon-on-insulator (SOI) technology, 65–67, 86, 87, 108
Simulation program with integrated circuit emphasis (SPICE), 53, 54, 57–59, 69, 83
Single-event functional interruption (SEFI), 35–36, 44
Single-event latchup (SEL), 40–44, 50, 67, 69, 71
Single-event transient (SET), 24, 30, 36–40, 42, 43, 57, 58, 70, 71, 73, 74, 81, 83–86, 88–110, 115–134
Single-event upset (SEU), 29–37, 39, 43, 50, 54, 65, 73–75, 86, 87, 95, 119
SOI technology, *see* Silicon-on-insulator (SOI) technology
Solar energetic particles (SEPs), 1, 2, 4, 25
SPICE, *see* Simulation program with integrated circuit emphasis (SPICE)
SRIM code, *see* Stopping and range of ions in matter (SRIM) code
Standard cell library, 81, 83, 87, 88, 96, 97, 116, 126
Standard-cell methodology, 81, 82, 85, 97, 115
Standard cells, 81, 82, 88, 96, 109, 115–120, 122, 126–132

STI oxide, *see* Shallow trench isolation (STI) oxide
Stopping and range of ions in matter (SRIM) code, 16–18, 56

T
TCAD simulation, *see* Technology computer-aided design (TCAD) simulation
Technology computer-aided design (TCAD) simulation, 23, 49, 50, 52, 70, 71, 95, 96, 99
Technology mapping, 116–118, 122–124, 133, 134
TF, *see* Transistor folding (TF)
TMR, *see* Triple modular redundancy (TMR)
Transistor folding (TF), 72, 94–110
Transistor sizing, 69, 70, 72, 77, 95, 97, 121
Transistor stacking (TS), 72, 82–94, 99, 109, 110
Triple modular redundancy (TMR), 73, 74, 132, 133
TS, *see* Transistor stacking (TS)

V
Van Allen belts, 2, 3, 5, 6, 25
Voltage variability, 104–106, 109